Smart Innovation, Systems and Technologies 18

For further volumes:
http://www.springer.com/series/8767

Robert J. Howlett, Bogdan Gabrys,
Katarzyna Musial-Gabrys, and Jim Roach (Eds.)

Innovation through Knowledge Transfer 2012

Editors
Prof. Robert J. Howlett
KES International
United Kingdom

Prof. Bogdan Gabrys
Smart Technology Research Centre
Computational Intelligence Research Group
Bournemouth University
Poole
United Kingdom

Dr. Katarzyna Musial-Gabrys
Department of Informatics
School of Natural and
Mathematical Sciences
Kings College
London
United Kingdom

Prof. Jim Roach
Dean, School of DEC
Bournemouth University (Talbot Campus)
Poole
United Kingdom

ISSN 2190-3018
e-ISSN 2190-3026
ISBN 978-3-642-43915-5
ISBN 978-3-642-34219-6 (eBook)
DOI 10.1007/978-3-642-34219-6
Springer Heidelberg New York Dordrecht London

Softcover reprint of the hardcover 1st edition 2013

Printed on acid-free paper

Springer is part of Springer Science+Business Media (www.springer.com)

InnovationKT-2012 Preface

This volume represents the proceedings of the Fourth International Conference on Innovation through Knowledge Transfer, InnovationKT-2012, organised jointly by KES International and the Institute of Knowledge Transfer, in partnership with Bournemouth University.

Featuring world-class invited speakers and contributions from a range of backgrounds and countries, the InnovationKT-2012 Conference was an excellent opportunity to disseminate, share and discuss the impact of innovation, knowledge sharing, enterprise and entrepreneurship.

The Innovation through Knowledge Transfer conference series provides an opportunity for researchers and practitioners in the field to publish their work. It also fosters the development of a community from the diverse range of individuals practicing various aspects of knowledge exchange and related disciplines.

InnovationKT-2012 called for both short papers and full papers which were reviewed by experts in the field and orally presented at the conference.

The conference featured a workshop (chaired by Prof. Bogdan Gabrys and Dr. Katarzyna Musial) arising out INFER, a major EU-funded project involving academic and industrial partners from three European countries (Germany, Poland and United Kingdom). The programme offered participants the opportunity to move between sectors and country in order to provide, absorb and implement new knowledge in a professional industrial-academic environment, a classic example of knowledge sharing in practice.

Thanks are due to the many people who worked towards making the conference a success. We would like to thank the Professor Jim Roach, Head of School of Design, Engineering and Computing at Bournemouth University, and Sir Brian Fender of the IKT for opening the conference. We would also like to thank the invited keynote speakers, the members of the International Programme Committee, and all others who contributed to the organisation of the event.

We hope you find InnovationKT'2012 an interesting, informative and useful event. We intend that future conferences in the InnovationKT series will continue to serve the knowledge transfer community and act as a focus for its development.

Robert J. Howlett
Bogdan Gabrys
InnovationKT'2012 General Chairs

Organisation

General Conference Chairs

Professor Bogdan Gabrys
Head of Smart Technology Research Centre, Bournemouth University, UK

Professor Robert J. Howlett
Executive Chair, KES International & Bournemouth University, UK

Honorary Conference Series Co-chairs

Sir Brian Fender CMG MInstKT
Chairman and President of the Institute of Knowledge Transfer

Mr. Iain Gray
Chief Executive, Technology Strategy Board

Institute of Knowledge Transfer Liaison Chair

Russell Hepworth Business Development Manager,
Institute of Knowledge Transfer

Innovation through Knowledge Transfer is organised and managed by **KES International** in partnership with **the Institute of Knowledge Transfer**.

International Programme Committee

Dr Mark Anderson	Glasgow Caledonian University, Scotland, UK
Dr Geoff Archer	Teeside University, UK
Ms Linda Baines	Rutherford Appleton Laboratory, UK
Prof. Francisco V. Cipolla-Ficarra	ALAIPO and AINCI, Spain / Italy
Mr Phil Cooper	National Physical Lab, UK
Dr Jarmila Davies	Wales Assembly Government, Wales, UK
Dr Paola Di Maio	University of Strathclyde, Scotland, UK
Ms Kim Dovell	Institute of Knowledge Transfer, UK
Ms Charlene Edwards	Melbourne, Australia
Sir Brian Fender	Institute of Knowledge Transfer, UK
Mr Eddie Friel	University of Ulster, Northern Ireland, UK
Dr Jonathan Gorst	Sheffield Hallam University, UK
Prof. Christos Grecos	University of the West of Scotland, Scotland, UK
Prof Ileana Hamburg	University of Applied Sciences Gelsenkirchen, Germany
Prof Owen Hanson	University of Middlesex, UK
Mr Russ Hepworth	Institute of Knowledge Transfer, UK
Prof. Robert Howlett	Bournemouth University, UK
Dr Jens Lønholdt	Technical University of Denmark, Denmark
Dr.-Ing. Maik Maurer	Technische Universität München, Germany
Prof Maurice Mulvenna	University of Ulster, Northern Ireland, UK
Dr Katarzyna Musial	King's College, London, UK
Prof Ian Oakes	University of Wolverhampton, UK
Prof Jim Roach	Bournemouth University, UK
Ms Val Wooff	Durham University, UK
Dr Cecilia Zanni-Merk	INSA-Strasbourg, France
Dr Shangming Zhou	Swansea University, Wales, UK

Dr Mark Anderson	Glasgow Caledonian University, Scotland, UK
Dr Geoff Archer	Teeside University, UK
Ms Linda Baines	Rutherford Appleton Laboratory, UK
Prof. Francisco V. Cipolla-Ficarra	ALAIPO and AINCI, Spain / Italy

Keynote Invited Speakers

Kelvin Pitman
Director of Open Innovation (recently retired)
Crown Packaging UL PLC

Working with Open Innovation Intermediaries

Abstract: Nearly ten years ago, Henry Chesbrough coined the term "Open Innovation". Even before it was given a name, many companies had been exploring what the concept meant for them. A key part of this has been the growth of both formal and informal networks to share and compare progress along the Open Innovation journey. Even more importantly, networks have been used to search for or offer technologies needed or being made available. Both types of network overlap, but in particularly the latter has relied heavily on the use of Intermediaries ranging in size from a single specialist consultants to some fairly large organisations. Kelvin Pitman, recently retired as Open Innovation Director for Crown Packaging Technology will outline a little of Crown's Open Innovation journey while case studying their use of an Intermediary. This will be illustrated by some of the academic work carried out in area mainly by Cambridge University Institute for Manufacturing in which Crown participated.

Biography

Kelvin Pitman recently retired as Director of Open Innovation for Crown working closely with Cambridge University Institute for Manufacturing, IXC, Industrial Research Institute (USA), NESTA and several other organisations to exploring how Crown Technology could exploit Open Innovation in a B2B situation. His career started after obtaining an honours degree in Polymer and Chemical Technology in 1974 when he joined Metalbox R&D in the UK. He progressed to Senior Scientist before taking on roles as Product Manager for beverage closures; then Commercial Office manager covering plastic closures for food, pharmaceuticals, automotive and household products. Subsequently he became Market Development Manager for PET bottles and beverage closures at divisional level. In 1988 he accepted a position as Licensing Manager for Crown Obrist, a subsidiary of Crown Holdings in Switzerland, expanding their beverage closure production technology to Belgium, Germany and the USA. Followed a period as Corporate Customer Manager he became Technical Director responsible for quality, R&D, production, customer technical service and purchasing. When Crown acquired Caurnaud Metalbox in 1996, Kelvin accepted the role of Global Closure Development Director, managing teams in the USA, UK and Switzerland. More recently, following a period as Director of Global Licensing, Kelvin took up his final role as Director of Open Innovation.

Caroline Bishop
Managing Director, IXC UK Ltd

Developing Smarter Engagements between HEIs and Corporate Business

Abstract: The InnovationXchange (IXC) has been putting the theory of Open Innovation into practice since its inception in 2004. Connecting technology demands in business with the supply of knowledge and expertise in HEIs and small companies, IXC is well placed to identify new trends and ideas in innovation practice and the use of open innovation paradigm in particular.

In this presentation Caroline Bishop, IXC's Managing Director will provide two case studies from Unilever and Siemens illustrating how corporate business is looking to build new business models around multi party consortia that will lever better, faster value from its engagement with HEIs. What corporate business has learnt about how to set up and manage these complex relationships will then be used to set out some of the key questions and actions HEIs will need to address if they are to engage successfully with this new corporate business agenda.

Biography

Following an early career with the NHS, Caroline subsequently spent 20 years at the University of Birmingham, in a variety of progressively challenging roles all focused on finding innovative new ways of engaging the University with the private and public sectors and raising income through knowledge exchange. Highlights include setting up the School of Public Policy's MBA programme in Hong Kong and building the University's business engagement strategy that saw the establishment of a team of business engagement staff that raised £32.5 million of new funding over the five years between 2001 and 2006. Especially important in this period was the creation of the Medici idea and the establishment in 2002 of a five-university partnership across the East and West Midlands to deliver training in technology transfer to medical scientists. Caroline subsequently led the bid that won £2 million from the Universities funding body (HEFCE) to train and develop medical scientists to engage more effectively in business. Medici was cited by HEFCE as an exemplar of innovation and good practice in 2007 and continues to provide service today. In 2006 Caroline secured £3.6 million of grant funding for the establishment of IXC UK and changed direction in her career to become Director of Operations for this new start up operation of the University of Birmingham. Six years on IXC UK has now spun out from the University and is in its fourth independent trading year and Caroline has become the company's Managing Director. IXC UK Ltd is designed to offer an alternative and more effective way to enable corporates, SMEs and Universities to collaborate and innovate. It is a very exciting arena in which to work and a huge challenge, which Caroline is relishing. Linking Jaguar LandRover with a small company in the Aerospace sector to enable Jaguar to develop a new hybrid engine based around gas turbines as opposed to the current industry approach with electric is a good example of what IXC enables its clients to do. The client base of IXC includes amongst others well known names such as, Crown Packaging, Jaguar Land Rover, Abbott Pharmaceuticals, Cadbury's, Smith and Nephew, Spriax Sacro, Kraft, PepsiCo, Dow Chemicals and MARS. Smaller companies such as Molecular Products, Phase Vision, Geotechnical Instruments, Cochlea and MiniFab also feature. The universities of La Trobe, Monash, Birmingham, Leeds, Manchester Metropolitan, Wolverhampton and Northumbria feature in the University client base whilst the Smith Family Foundation exemplify work in the social enterprise field and public sector.

Dr Piero P.Bonissone
Chief Scientist, Coolidge Fellow
General Electric Global Research, USA

Soft Computing in Prognostics and Health Management (PHM) Applications

Abstract: Soft Computing (SC) is a term that has evolved, since its inception in 1991, to represent a methodology and a set of techniques covering the aspects of data-driven models design, domain knowledge integration, model generation, and model tuning. We distinguish between offline Meta-heuristics (MH's), used for model design and tuning, and online MH's, used for models selection or aggregation. This view suggests the use of hybrid SC at each MH's level as well as at the object level. We manage model complexity by finding the best model architecture to support problem decomposition, generate local models with high- performance in focused applicability regions, provide smooth interpolations among local models, and increase robustness to imperfect data by aggregating diverse models.

Within the broad spectrum of Soft Computing (SC) applications, we will focus on Prognostics & Health Management (PHM) for assets such as locomotives, medical scanners, aircraft engines, etc. The main goal of PHM is to maintain these assets' operational performance over time, improving their utilization while minimizing their maintenance cost. This tradeoff is typical of long-term service agreements offered by OEM's to their valued customers. Typical PHM functions range from anomaly detection, to anomaly identification, failure mode analysis (diagnostics), estimation of remaining useful life (prognostics), on- board control, and off board logistics actions.

We illustrate this concept with a case study in anomaly detection for a fleet of physical assets (such as an aircraft engines or a gas turbines.) Anomaly detection typically uses unsupervised learning techniques to extract the underlying structural information from the data, define normal structures and regions, and identify departures from such regions. We focus on one of the most common causes for anomalies: the inadequate accuracy of the anomaly detection models, which are prone to create false alarms. To address this issue, we propose a hybrid approach based on a fuzzy supervisory system and an ensemble of locally trained auto associative neural networks (AANN's.) The design and tuning of this hierarchical model is performed using evolutionary algorithms. In our approach we interpolate among the outputs of the local models (AANN's) to assure smoothness in operating regime transition and provide continuous condition monitoring to the system. Experiments on simulated data from a high bypass, turbofan aircraft engine model demonstrated promising results.

Biography

A Chief Scientist at GE Global Research, Dr. Bonissone has been a pioneer in the field of fuzzy logic, AI, soft computing, and approximate reasoning systems applications since 1979. His current interests are the development of multi-criteria decision making systems for PHM and the automation of intelligent systems lifecycle to create, deploy, and maintain SC-based systems, providing customized performance while adapting to avoid obsolescence. He is a Fellow of the Institute of Electrical and Electronics Engineers (IEEE), of the Association for the Advancement of Artificial Intelligence (AAAI), of the International Fuzzy Systems Association (IFSA), and a Coolidge Fellow at GE Global Research. He is the recipient of the 2012

Fuzzy Systems Pioneer Award from the IEEE Computational Intelligence Society. Since 2010, he is the President of the Scientific Committee of the European Centre of Soft Computing. In 2008 he received the II Cajastur International Prize for Soft Computing from the European Centre of Soft Computing. In 2005 he received the Meritorious Service Award from the IEEE Computational Intelligence Society. He has received two Dushman Awards from GE Global Research. He served as Editor in Chief of the International Journal of Approximate Reasoning for 13 years. He is in the editorial board of five technical journals and is Editor-at-Large of the IEEE Computational Intelligence Magazine. He has co-edited six books and has over 150 publications in refereed journals, book chapters, and conference proceedings, with an H-Index of 27 (by Google Scholar). He received 61 patents issued from the US Patent Office (plus 50 pending patents). From 1982 until 2005 he has been an Adjunct Professor at Rensselaer Polytechnic Institute, in Troy NY, where he has supervised 5 PhD theses and 33 Master theses. He has co-chaired 12 scientific conferences and symposia focused on Multi-Criteria Decision-Making, Fuzzy sets, Diagnostics, Prognostics, and Uncertainty Management in AI. Dr. Bonissone is very active in the IEEE, where is has been a member of the Fellow Evaluation Committee from 2007 to 2009. In 2002, while serving as President of the IEEE Neural Networks Society (now CIS) he was also a member of the IEEE Technical Board Activities (TAB). He has been an Executive Committee member of NNC/NNS/CIS society since 1993 and an IEEE CIS Distinguished Lecturer since 2004.

Mike Smith
Pro-Vice Chancellor Research
Sheffield Hallam University, UK

Co-creation of Innovation

Abstract: This presentation will propose that the concept of the co-creation of innovation will be an important paradigm for the development of innovation in the future. In the recent past there have been debate about whether knowledge exchange is a more appropriate descriptor than knowledge transfer, for the activities undertaken by organisations such as Universities which contribute to the innovation agenda. It will be argued that the requirement for innovation to have a greater impact on economic transformation in the short, medium and long-term, requires an improvement in the quality and relevance of innovation, which may not be able to be delivered by either knowledge transfer or knowledge exchange alone.

The value of multidisciplinary collaboration in research is widely recognised, and the concept of the co-creation of innovation is an extension of the principle to the joint development of innovation between universities and the commercial sector. An improvement in the quality and relevance of the innovation will increase the likelihood of its implementation and, as a consequence, its impact. Mechanisms to achieve co-creation of innovation will be discussed in the context of different international models.

The importance of trust, and its associated influence on risk, will be discussed and a model will be proposed which describes the route-map for the co-creation of innovation between public sector organisations such as Universities or the NHS and the commercial sector. A number of examples will be presented from the author's experience including product development and the provision of higher skills training in areas such as advanced manufacturing, healthcare and business and management. Implications for Universities will be discussed, and the barriers to progress will be outlined.

Biography

Professor Mike Smith has a long experience of working in the University, NHS and commercial sectors and has a wide network of international links. Among his previous posts, he was dean of research for the Faculty of Medicine, Dentistry, Psychology and Health at the University of Leeds; director of R&D for the Leeds Teaching Hospitals Trust; chair of medical physics and head of the University and NHS Departments of Medical Physics at the University of Leeds and Leeds Acute Hospitals; director of medical physics and clinical engineering at Guys Hospital and the University of London as well as being medical scientist at the University of Edinburgh.

Mike also has a long track record in research, supporting educational developments as well as commercialising research and innovations. This includes publishing patents, licensing research developments, forming start-up and spin-out companies and creating private sector vehicles to commercialise research output.

He holds a number of non-executive directorships in the commercial and NHS sectors. In addition, Mike has over 20 years experience of serving on major professional and governmental national committees, contributing at a national level on policy and funding issues. Examples include the HEFCE Research Assessment Exercise, MRC Medical Advisory Board, NHS Central Research and Development Committee, DTI Foresight Taskforces on Biotechnology, Pharmaceuticals and Medical Devices and president of the British Institute of Radiology in its Centenary Year. While Deputy Vice-Chancellor, Research and Enterprise at the University of Teesside, he has played a major role, working with the Regional Development Agency in putting together the Digital City Project and securing a £10m building for the University to support this initiative.

Anne Snowdon
Professor, Management Science
Odette School of Business, University of Windsor, Canada
Chair, International Centre for Health Innovation
Richard Ivey School of Business

Issues in International Health Innovation

Abstract: Innovation is widely viewed in many sectors as a strategy for stimulating economic growth and prosperity in countries around the world. In particular, knowledge-intensive sectors such as finance, manufacturing, information systems and automotive all rely on innovation to remain viable and competitive in global markets. However, innovation has been slow and limited in achieving similar gains in productivity, efficiency and sustainability in the health sector. This presentation will examine the unique challenges health systems experience in achieving innovation adoption. Evidence of the importance of innovation adoption in health care will be presented through the profiles of developed countries globally. Opportunities to stimulate innovation adoption in health systems using an integrated knowledge translation approach will be profiled and discussed, including the key factors that support innovation adoption in health care system. Health innovation projects that use an integrated knowledge translation approach will be presented, including opportunities for using this approach to engage key stakeholders such as health professionals, health consumers, private sector partners and health system leaders in driving innovation adoption in health systems.

Biography

Dr. Anne Snowdon is Chair of the International Centre for Health Innovation at the Richard Ivey School of Business. Located within The University of Western Ontario in Canada, the Centre's mandate is to build health system leadership capacity to support and drive innovation adoption of technologies, systems and processes that our health systems need in order to be sustainable.

Dr. Snowdon is an Adjunct Faculty member at Ivey and holds an Adjunct appointment in the School of Nursing at McGill University in Montreal. She is also a Professor at the Odette School of Business and is cross-appointed to the Faculty of Engineering at the University of Windsor.

Dr. Snowdon holds a Bachelor of Science in Nursing from The University of Western Ontario, a Master of Science from McGill University and a PhD in Nursing from the University of Michigan. She is a Fulbright Scholar and was awarded the Social Sciences and Humanities Research Council Doctoral Fellowship for her research on parenting during childhood hospitalization. Dr. Snowdon is also the Theme Coordinator for Automotive Health and Safety for Canada's automotive research program,

AUTO21 Network of Centres of Excellence, and has commercialized innovative new safety seat products for occupants in vehicles as a result of this research program.

In addition to her expertise in health system leadership and innovation, Dr. Snowdon's research also looks at the role of engaged consumers as agents of change and reform to health systems.

Jim Roach
Professor & Dean, School of Design, Engineering and Computing
Bournemouth University

Industry Engagement: Barriers and How to Overcome Them

Abstract: This presentation covers the research we have done to identify the problems industry face when trying to interact with the University. The findings of focus groups and face to face interviews with companies that have an existing relationship with the University and those companies that have no relationship with the University but would like to engage. Discussion will enquire what sort of interactions does the University want and how is it promoting them. Then from our findings we propose a "tool kit" for Universities to build successful relationships with Industry.

Biography

Professor Jim Roach took up his post of Dean of the School of Design, Engineering and Computing in September 2008 having previously served as Deputy Dean (Education) in the School.

Professor Roach has great experience of working with Industry and Commerce linking with Academia. In 2001 Professor Roach was awarded one of only twelve National Business Fellowships in recognition of the collaborative work undertaken between the University and Industry. Through his work with Industry over twenty collaborative programmes (Teaching Company Schemes now Knowledge Transfer Partnerships) have been carried out. These have resulted in many new products and improved processes for local companies. Research interests are in simulation and modelling of electronic circuits, programmable electronics, electronic product generation, protecting Design Rights and Intellectual Property and the management and development of engineering innovation. Currently Professor Roach is collaborating with a local company developing an intelligent temperature monitoring instrument for use in the power industry.

Professor Roach has a strong connection with the Royal School of Signals, Blandford and was instrumental in obtaining accreditation for a number of military courses allowing them to attract civilian qualifications. The MSc in Communications and Information Systems Management has run since 1999 and this has been followed by a BSc(Hons) Telecommunication Systems Engineering and most recently a FdSc Management of Military Information Systems and a FdSc in Communications Systems engineering

Contents

INFER Project: Challenges in Transfer of Knowledge to Industry Chairs: Prof Bogdan Gabrys (Bournemouth University) and Dr Katarzyna Musial (Kings College London)

Next - Practice in University Research Based Open Innovation Chair: Jens Rønnow Lønholdt (Technical University of Denmark)

Social Innovation and Related Paradigms Chair: Jarmilla Davies (Wales Assembly Government)

Understanding the Other Side – The Inside Story of the INFER Project

Katarzyna Musial[1], Marcin Budka[2], and Wieslaw Blysz[3]

[1] King's College London, School of Natural & Mathematical Sciences, Department of Informatics, United Kingdom
katarzyna.musial@kcl.ac.uk
[2] Bournemouth University, Smart Technology Research Centre, United Kingdom
mbudka@bournemouth.ac.uk
[3] Research & Engineering Center, Poland
wieslaw.blysz@rec-global.com

Abstract. In the last few years, the collaboration between research institutions and industry has become a well established process. Transfer of Knowledge (ToK) is required to accelerate the development of both sides and to enable them to unlock their full potential. European Commission within the Marie Curie Industry and Academia Partnerships & Pathways (IAPP) programme supports the cooperation between these two sectors at the international scale by funding research projects that as one of the objectives aim at enhancing human mobility. IAPP projects offer people from different institutions the possibility to move sector and country in order to provide, absorb and implement new knowledge in a professional industrial-academic environment. In this paper, one of such projects is presented and both academia and industry perspectives in regard to opportunities and challenges in Transfer of Knowledge are described. Computational Intelligence Platform for Evolving and Robust Predictive Systems (INFER)[1] is the IAPP project that serves as a case study for this paper.

1 Introduction

Transfer of Knowledge (ToK) is one of the concepts that are self-explanatory. It is transferring the knowledge from one entity (person, group or a whole institution) to another one.

Some of the definitions of ToK can be found on the websites of Knowledge Transfer Centres. For example, Knowledge Transfer Centre from University of St Andrews defines ToK as "the systems and process by which research institutions interact with businesses, the public and other organisations to enable knowledge and expertise to be utilised leading to innovative, profitable and social improvements" (Knowledge Transfer Centre from University of St Andrews, 2011).

---

[1] For more information please see http://www.infer.eu

R.J. Howlett et al. (Eds.): *Innovation through Knowledge Transfer 2012*, SIST 18, pp. 1–9.
DOI: 10.1007/978-3-642-34219-6_1

Innovation through Knowledge Transfer organisation provides more detailed explanation of what Knowledge Transfer is from their perspective: "We define knowledge transfer as the means by which expertise, knowledge, skills and capabilities are transferred from the knowledge-base (for example, a university or college, a research centre or a research technology organisation) to those in need of that knowledge (for example a company, social enterprise or not-for-profit organisation). Hence, knowledge transfer involves the interface between universities and business, and involves the commercialisation of skills and expertise possessed by higher education. The purpose of knowledge transfer is to catalyse and facilitate innovation" (InnovationKT, 2011)".

Another definition was proposed by the Knowledge Transfer Centre from Queens University Belfast: "Knowledge Transfer Partnerships are based on partnerships between academic groups and companies who need access to skills and knowledge in order to innovate."[2]

Note, that all of the definitions point out that the purpose of Knowledge Transfer is to facilitate innovation. ToK is a very challenging task as the knowledge is not tangible and usually defragmented across the organisation. The organisational knowledge is tactic or hard to articulate (Nonaka and Takeuch, 1995). In addition, there are a lot of sources of knowledge, e.g. people employed in the organisation, tools used across the company, groups within the organisation and tasks assigned to people (Argote and Ingram, 2000).

As it will be presented, the perception and importance of ToK process is different for people with different backgrounds, positions in the company, and from different sectors. Author's experience shows that factors influencing the success of Knowledge Transfer are among others: (i) commitment level of both top management and teams involved in project, (ii) resources available for ToK activities or (iii) type and quality of prepared ToK activities.

The motivation to write this paper was to make aware people who want to undertake large projects where ToK activities are one of the main focuses that the ToK process can be well implemented but it requires a lot of planning and preparation. ToK can be successful only if the working environment is set up in a way that sites understand and respect each other. It should be emphasized that people are inseparable part of the transfer of knowledge. They should be made aware of the importance of ToK activities. The whole process will perform smoothly if people are committed and know what is expected from them and what they can expect from other team members.

This is experience-based study prepared by people who were and are actively involved in the Knowledge Transfer in the INFER project. First part of the article presents the INFER project and different types of Transfer of Knowledge activities that are implemented as part of the work programme. The next part is devoted to describe the industry and research perspectives of people participating in the project. Finally, both the opportunities arising from Knowledge Transfer and challenges that need to be overcome to enable successful ToK are outlined and discussed.

---

[2] There exists many more Transfer of Knowledge Centres. The presented definitions aim at providing general idea about ToK.

2 INFER Project

As it was mentioned before, INFER stands for Computational Intelligence Platform for Evolving and Robust Predictive Systems and is a project funded by the European Commission (EC) within the Marie Curie Industry and Academia Partnerships & Pathways (IAPP) programme, with a runtime from July 2010 until June 2014. The programme will offer involved people the possibility to move sector and country in order to provide, absorb and implement new knowledge in a professional industrial-academic environment.

INFER involves employees from organisations from three different countries. This includes Evonik Industries from Germany, one of the world's leading companies in the process industry; Research & Engineering Center (REC) from Poland, a highly innovative software engineering company and the Smart Technology Research Centre of Bournemouth University (BU) in the UK, an interdisciplinary and integrative centre conducting research in the field of automated intelligent technologies.

The area of the project is pervasively adaptive software systems for the development of an open modular platform applicable in various commercial settings and industries. The main innovation of the project is a novel type of environment in which the "fittest" predictive model for whatever purpose will emerge – either autonomously or by user high-level goal-related assistance and feedback. Such system is beneficial for businesses relying on accurate ahead predictions of any type (e.g. customer behaviour, market conditions) and, at the same time, requiring an automated ability to react to changes in market or operational conditions. As the project is funded by EC under the IAPP programme one of its main objectives is the continuous knowledge transfer.

During the preparation of the project proposal all parties discussed the expertise that each partner brings to the project and how others can benefit from this knowledge. The identified transfer of knowledge opportunities were presented in the proposal from two perspectives: industry and academia.

a) **The industrial partners (REC nad Evonik)** can benefit from the latest research developments deriving from the academic sector and helps to ensure their fast and cost-effective validation and application at industrial scale. Early Stage Researchers (ESR) and Experienced Researchers (ER) from industry have the opportunity to gain knowledge about innovative approaches and to build the basis for a successful employment of these methods in practice.
b) **The academic partner (BU)** has the opportunity to (i) gain experience in an industrial environment, (ii) adapt and prove their concepts, (iii) develop more expertise in real industrial application and (iv) gain knowledge and experience in organising, realisation and managing complex, multi-site and multi-vendor projects as well as transferring successful research into commercial software products.

In order to be able to realise those identified opportunities the detailed plan of activities that would be undertaken was needed. The set of ToK activities that was planned as a part of the INFER project are presented in the next section.

3 Transfer of Knowledge Activities

The activities connected with the ToK concept are organised around: (i) providing, (ii) absorbing and (iii) implementing new knowledge in an organization. ToK activities can be divided into internal and external events. The former are those where only people involved in the project are included and they are usually more focused on a specific topic directly connected with the developed project. The external actions aim to share knowledge and experience at the bigger scale and because of that they tend to be more general but equally valid as the internal ones. Common activities that are associated with Transfer of Knowledge are (i) conferences and workshops, (ii) trainings and tutorials, (iii) discussions and brainstorming sessions, etc. A very interesting mechanism of a growing importance is the idea of *secondments* that enables people to work at the partner site and to gather knowledge about what and how the processes are used by the partner as well as to gain understanding about partner's know-how.

There is also a part of Transfer of Knowledge that is not and cannot be done by means of planned and organised events. For example a brief chat with a person from another sector about work-related issues can be seen as the Knowledge Transfer activity. This type of activity is not planned in advance. However, what should be emphasized, the environment that enables these two people to meet has to be set up beforehand.

The mechanism implemented in the project that to a great extent facilitates ToK is the *secondment* of staff. All research activities and platform implementation were organised and schedule around the secondment plan.

Transfer of Knowledge activities that take place within the INFER project for both project participants and also external community are:

a) **Presentations of seconded staff** – a mechanism that supports exchanging knowledge and experience. All researchers who go for the secondments need to prepare three presentations: (i) upon arrival at the host organisation: on the latest developments at their home organisations, (ii) at the end of their stay: on the progress made during their stays, and (iii) upon their return to home organisation: on knowledge gained during their stays at the host organisations.
b) **Tailored short courses** that are delivered by more experienced researchers from each partner (for the list of short courses within INFER procect see Table 1).
c) **Invited lectures, workshops, and conferences** that enable to present the project to external community. For example workshop on Smart Adaptive Systems that is organised annually by BU.
d) **Meetings with all partners** (and all researchers involved) organised once a year in order to discuss progress made and to exchange ideas for future activities.
e) **Activities that are part of the standard educational and staff development programs of the host organisations** are made available to the seconded researchers.
f) **Both formal and informal research meetings, discussions etc.**

Table 1 Tailored short courses planned within the INFER project

Who delivers?	To whom?	Course Title
BU	Evonik, REC	Advanced research methods
BU	Evonik, REC	Multiple classifier and prediction sys
BU	Evonik, REC	Adaptive evolving structures for continuous prediction
BU	Evonik, REC	Nature-inspired data mining
BU	Evonik, REC	Meta learning and complex systems
BU	REC	Advanced software engineering practices
Evonik	BU	Model-driven soft sensors
Evonik	BU	Managing large industrial software projects
Evonik	BU	Process automation strategies
Evonik	BU	Advanced control in chemical processes
REC	BU	System and functional tests
REC	BU	Development of commercial projects
REC	BU	Agile project management

All these formally defined mechanisms aim really at facilitating people to exchange their expertise, experience and knowledge. A very important element of all these meetings, presentations and conferences is to gather people in one room, help them to know each other and engage them with a project in a way that each person will find something of interest for him or her.

4 Perception of Transfer of Knowledge

4.1 Academic Perspective

The research, which takes place in the academic environment, is a very specific activity. On one hand it tackles challenging, unresolved issues with a potentially very significant impact on the society, economy, environment etc. On the other hand, the results of that research can often be difficult if not impossible to exploit commercially or for other applicable purposes. This is because the researchers often embark on tasks, which after investing a lot of effort, turn out to be impossible to realise. Hitting dead ends is a natural cost of progress as it seems. It is difficult not to agree with one of the famous Einstein's statements: "If we knew what it was we were doing, it would not be called research, would it?"

A somewhat simplistic but still useful perspective of the differences between academia and industry comes down to the expected outputs or deliverables. While in the industrial setting the deliverables are typically products with strictly specified properties and delivered at a specified time, in academia a deliverable is usually a publication. Yes, it needs to meet some quality standards, which depend on the requirements of the journal or conference the publication is targeting, and yes, there are usually also some time constraints (e.g. an academic is required to produce a given number of publications per year), but the process of producing the output is nothing like the processes implemented in the industry. As the result, from the academic perspective, the processes, which allow for efficient realization of the day to day activities of the industrial partners, are of great interest. But how can this type of knowledge be transferred in an efficient way, along other activities (research, software development) within the INFER project?

We believe that the answer is the "learn by doing" paradigm. There is no better way for familiarizing an academic with the project management or software development and testing processes as well as tools which facilitate them, than experiencing the processes from the inside. And this can be achieved by making the academic a member of the industrial partner's team dedicated to the project, i.e. by the means of secondments which enable a two-way knowledge transfer by means of direct interactions. These activities are also of interest for the industry. To what extent the knowledge acquired this way can actually be implemented in the academic environment is a different issue. Although as argued above, research is not a strict process with a guaranteed outcome, we believe that it can and should be managed to some extent, just like any other project.

Other transfer of knowledge activities, as organising lectures, workshops, conferences are the regular activities that researchers perform, sometimes without thinking about them in terms of Knowledge Transfer. These things just happen on the daily basis.

4.2 Industry Perspective

The perception of various Transfer of Knowledge activities in the industry heavily depends on two factors: the backgrounds and the positions of the people involved. While it is natural that the expectations of top and middle management are quite different to those of the engineering staff, it forces the ToK activities to be performed at different levels and with different scopes. For example, although general, high-level presentations and lectures can be of interest to people without engineering background, they are usually perceived by the development team as an event that is not relevant for them. Hence, in practise the ToK activities within INFER are directed at either the management and sales staff or the software development team.

The activities prepared for the first group are mainly presentations that focus on the benefits of predictive analytics and unique features of the INFER platform being developed, potential applications of the platform as well as the portfolio of successful past projects and collaborations of the academic members of the INFER team. This audience is mostly interested in gaining knowledge about the project and participant's expertise, which enables them to professionally offer the INFER platform and accompanying services (trainings, consultancy etc.) to potential customers. The experience shows that presentations and seminars are probably the most efficient way of delivering this kind of knowledge.

Transfer of Knowledge between the scientific team and the software engineers involved in the project has mainly an informal character and its main elements are direct interactions between individuals or small groups of people. ToK is usually triggered by a need to make design and implementation decisions. In such situation both scientists and software engineers discuss the potential solutions and have chance to learn from each others. The secondments which take place within the INFER project are the key factor enabling these face to face interactions.

5 "Why Do We Need Each Other?"

There is a substantial and multidimensional benefit from cooperation and knowledge transfer between academia and industry. It comes from natural difference in approach that both parties naturally adopt for technology exploration, problem solving and making new technical ideas implemented and useful.

Industry tends to explore new technologies only in the situation where it has isolated a defined problem that it wants to solve and there is no known, suitable technology that has been already adopted by this industry. This gives an optimal resource utilisation and solution costs, at least locally. However, often the chosen approach and results are sub-optimal due to the focus on reaching the local goal, but not finding a shortest, cheapest and repeatable way to address the stated problem.

Academia often explores new technologies for finding new ways of using it (but still not really applying it) or even finding the next steps of evolution that would be applicable for a given technology. This is also done without a direct application of economic aspects for the experimentation phase. As a result, more optimal technical solution can be worked out that can be more optimal globally than a solution worked out by the industry itself.

A combination of both approaches can result in better and more rapid technology development and commercialisation. Also a big advantage of involving both parties (academia and industry) into a project is the fact that the combination of both approaches may result in higher number of disruptive technologies being developed and commercialised. This may lead to winning position in technology advancement, which is an important and common goal of every economy in the world. The cooperation and Transfer of Knowledge between academia and industry can help both sides to reach their goals and identify those that are common. As an example it should be mentioned that an invention that has been made by an academic has totally different value and recognition when it is effectively commercialised and vice versa, a top-notch invention generates often a new quality in business of the industry.

A natural development path of a new useful invention is its implementation and commercialisation. Thanks to Transfer of Knowledge between academia and industry there is a better understanding between both sides in the areas of: what are the most compelling problems of the industry that can be addressed by the academia, what are the best findings and inventions that are there to be applied, what are the true criteria that must be fulfilled to make a new technology applied, commercialised and thus useful.

There are also more positive "side effects" of Transfer of Knowledge and cooperation and these are: (i) transfer of knowledge related to the needs of the industry with regard to education of new workforce, (ii) networking that may result in new matches of client-vendor and thus new business opportunities can be generated and explored.

6 Transfer of Knowledge Challenges

There are some critical factors that should be considered before any transfer of knowledge activity takes place. The elements that have to be considered are:

- identifying and agreeing the goals and benefits of ToK between partners;
- planning the resources (needed and available);
- establishing an approach to ToK and means of communication;
- team building;
- getting attention;
- execution with quality, gathering and providing feedback.

First, the goals and potential benefits of transfer of knowledge activities should be identified by each of the ToK partners, clearly communicated and agreed.

Secondly, the appropriate resources should be planned and provided. This gives a good chance that ToK activity will be accomplished and the planned results will be achieved. Also, the communication should be aligned, as ToK activity is being performed between quite different organisations and environments. This requires a good mechanism implemented that will make all the parties coming together on communication level. One of the mechanisms would be to ensure social activities that will make different parties coming together – this is a critical factor to support clearing up problems in understanding major differences in approach, style of work and goals of all parties.

Fourth factor is the team building. It is really challenging to build an interdisciplinary team of experts and researchers who come from (and still are a part of) different organisations. In order to reach that there should be a mechanism that ensures the team leader has the abilities and skills for effective team building.

Fifth factor is to attract attention of other parts of organisations that may benefit from ToK. Some mechanisms are needed here as well. We used internal publications to address that. We plan to have some more events for ToK inside of organisations and these include e.g. presentation to other departments that can be potentially interested in the project outcome.

Sixth factor is to execute with quality, gather and provide feedback to people involved in ToK to help them improve on ToK activities.

Each of these factors can be seen as a challenge in the ToK process. Taking into account different perspectives on Knowledge Transfer, the importance of common understanding cannot be neglected.

7 Conclusions

Transfer of Knowledge is a very important part of the INFER project. It requires preparing the plan of the process and activities involved in it and at the same time addressing the challenges presented above. It should be noted that in order to do this, both academia and industry perspectives have to be taken into account. Neglecting one of the parties can have disastrous effect. The element that is equally important as formal process is the working environment in which the defined procedures are implemented. In extreme situation lack of common understanding together with lack of willingness of partners to commit time can result in failing to implement planned ToK activities.

From our experience the secondment mechanism facilitates the ToK as it enables people with different backgrounds (engineers and researchers) to work together towards a specific goal that could not be achieved if only one group worked on it. Thus, we believe that Knowledge Transfer enhances innovation and enables the emergence of collective intelligence phenomena.

References

Argote, L., Ingram, P.: Knowledge transfer: A Basis for Competitive Advantage in Firms. Organizational Behavior and Human Decision Processes 82(1), 150–169 (2000)

InnovationKT, `http://innovationkt.org/` (last accessed November 25, 2011)

Knowledge Transfer Centre from University of St Andrews, `http://www.st-andrews.ac.uk/ktc/` (last accessed 25, 2011)

Knowledge Transfer Centre from Queen's University Belfast, `http://www.qub.ac.uk/directorates/KnowledgeTransferCentre/` (last accessed November 25, 2011)

Nonaka, I., Takeuchi, H.: The Knowledge-Creating Company. Oxford University Press, New York (1995)

Smart Meetings: Experimenting with Space

Jeremy Frey, Colin Bird, and Cerys Willoughby

Chemistry, University of Southampton, University Road, Southampton SO17 1BJ
{J.G.Frey,colinl.bird,Cerys.Willoughby}@soton.ac.uk

Abstract. During May and June 2011, motivated by the need to improve techniques for recording the processes and outputs of research, we ran two workshops under the auspices of the e-Science Institute. The theme title was "Smart Spaces for Smart People". Although our initial intention was to explore interactions between the physical and digital worlds, the emphasis changed to the productive exploitation of spaces ascribed as *smart*. We explored the quality of *smartness* in the context of *smart meetings*, which led us to conclude that the role of hardware and software technologies is to *confer capability*. For a *system* to achieve smartness, we deem certain components to be essential, most notably people. However, we also consider the role of both technological and traditional techniques for capturing meeting outcomes. We learned lessons that are applicable not only to meetings about research but also in the more general knowledge transfer context. We conclude that the way forward for exploiting smart spaces relies on design and on empowering the users of such spaces in that design. This paper is the first in a series of three, each dealing with different aspects of the workshops and how they influenced our thinking about knowledge transfer meetings, particularly in the context of sharing research outputs.

1 Introduction

One influential outcome of the e-Science and e-Research programmes that ran from the year 2000 onwards was a perceived need for improved and more dynamic techniques for recording the processes and outputs of research. New technologies, for example sensors and mobile devices, were developing at the same time, and some of the more mature technologies were becoming more sophisticated. Interest grew in exploring novel methods for recording both procedures and data in the context of the research or teaching activity.

We obtained funding from the e-Science Institute (eSI) to investigate interactions between physical spaces and personal digital technologies, and look for innovative ways to 'mashup' the two worlds. We established the project as a mini-theme entitled "Smart Spaces for Smart People", intending to consider smart environments in general, but with specific attention to research, teaching, and meeting spaces. As was the practice for other eSI themes, we organised facilitated workshops that had specific objectives.

This paper describes the organisation of the workshops and the overall experience so running them. We wanted our range of recording methods to be

R.J. Howlett et al. (Eds.): *Innovation through Knowledge Transfer 2012*, SIST 18, pp. 11–17.
DOI: 10.1007/978-3-642-34219-6_2

complementary while allowing a comparison of their relative merits. Our intention was always to explore interactions in various forms, trying to make the supporting technology as unobtrusive as possible. Reviewing the published literature about smart spaces reveals a strong tendency to describe the enabling technology, so our focus on interaction was to some extent novel, but we believe that knowledge transfer depends much more on human interaction than on technology. We explore that point in greater depth in the second paper of the series [5] and consider how best to apply our workshop experiences to knowledge transfer in the third paper [1].

When reviewing the first of the two workshops, it became apparent that the emphasis had changed from the anticipated exploration of interactions between the physical and digital worlds to the productive exploitation of spaces ascribed as *smart*. These changes came about from the discussions in the first workshop and from considering issues that came up when assessing the detailed technical aspects of holding the workshops in different venues. Although much of the focus was on *smart meetings*, we remained aware of the continuing need to consider other environments, such as learning and research. We developed four sub-themes and three key considerations as a basis for the successful planning and conduct of smart meetings. In our paper about exploiting smart meetings for knowledge transfer [1] we examine the principal ideas associated with each sub-theme, and some of the questions arising from the three considerations, and propose strategies for the effective utilization of smart meetings for knowledge transfer.

2 Workshop Methodology

Themes organized under the auspices of the e-Science Institute [2] focus on a specific issue in e-Science that crosses boundaries and raises new research questions. In our Theme description [3], we stated our objectives broadly as follows:

- To investigate the interaction between the use of Smart Spaces in the physical world and smart personal systems both technological and software;
- To explore and define best practice in enhancing the utility of the link between the physical and digital worlds.

We ran two workshops, the first at the e-Science Institute, Edinburgh, in May 2011, and the second at the University of Southampton, in June 2011. As we shall discuss later, the differences between the two venues provided valuable evidence regarding the influence of the space itself on the extent to which it can be ascribed as *smart*.

We began the first workshop very much with an open mind, so invited participants with a range of interests, albeit constrained by availability at what was fairly short notice. Similarly, although motivated more by a desire to make a fresh start to exploring the potential impact of *smart interactions*, we did not do a literature survey beforehand. With hindsight, that break with the past probably assisted both workshops to focus on the exploitation of smart interactions rather than the smart technology itself.

In the opening session of each workshop, we asked participants to introduce themselves briefly, and to focus on how they perceived the future for Smart Spaces, adding that ideas were what we were looking for. Complying with the original objectives of the Theme, we considered the workshops themselves to be experiments in using smart technologies (although we did not say so openly at the outset).

Rather than attempting to monitor interactions between equipment (subsequently termed instrumentation) and people, our attention was mainly on recording methods. In particular, we did not deploy any embedded sensors. The meeting rooms for both workshops contained audio-recording devices and offered wireless access to Twitter: we used the tag: *#smartspaces*. For the first workshop, we also used video recording and had access to the eSI Theme Wiki. Note-takers kept written records of each session, focusing on capturing key points, ideas, and remarks that redirected the discussion. For the second workshop, we had neither video recording nor an official note-taker, although one participant did make notes and provided a copy afterwards. For the second workshop, we also projected the Twitter feed onto a screen.

The same person (CB) acted as facilitator for both workshops, using flipcharts to capture the key points raised during proceedings. Subsequently, we assessed these traditional capture methods, notes and flipcharts, for both intrinsic usefulness and complementarity with the other recording methods.

It was during the planning stage for the first workshop that using the workshop itself as an experiment in running smart meetings emerged as a meta-objective. Technical limitations also became apparent at this stage. Equipment that we had initially envisaged using to support the workshops had performance limitations and long lead times and other options that might have offered satisfactory solutions were significantly more expensive than would have been reasonable in aid of a comparatively small workshop.

3 Results and Discussion

During the final sessions of the first workshop, it became very apparent that the emphasis had changed from the anticipated exploration of interactions between the physical and digital worlds to the productive exploitation of spaces ascribed as *smart*. Moreover, despite our continuing awareness of the need to consider other potentially smart environments, the workshop participants had focused very much on meeting spaces. The second workshop continued in the same vein, albeit concentrating on different aspects of meetings and the meaning of *smartness* in the context of meetings.

Reflecting the shift in our discussions predominantly towards how we might exploit *smart* spaces to enhance the conduct of meetings, the following additional goals (which we came to regard as meta-objectives) came into play:

- To gather requirements for a *smart meeting log system*, and to prepare a draft specification for the specialist supporting software required for such a log system.
- To evaluate critically the influence on the success of a meeting of the following tactics: using a facilitator; having the facilitator keep a visual record (on

flipcharts, for example); using an independent note-taker; maintaining an audio and/or video record of the proceedings; and emphasising discussion over prepared presentations.
- To investigate voice-to-text transcription, giving particular attention to: (a) whether individual participants can be identified from a single track; and (b) investigating the most useful and appropriate methods for searching and tagging the transcribed text, with a view to ensuring effective cross-linking with the meeting log.

Shortly after the second workshop, we began a survey of the literature relating to previous smart spaces work. Although we hope in due course to publish the results of that survey in the form of a review, we include in this paper some pertinent reflections arising from the survey, because they inform our discussion of the outcomes from the two workshops. Our observations are as follows:

- The smart spaces paradigm (or meme) emerged as a result of advances in ubiquitous computing, also known as pervasive computing. To some extent, this device-centric view accounts for the dominance of environments with capabilities driven by the technology available, some of which is embedded.
- The word *context* appears in most recent publications about smart spaces, but almost all authors interpret the term as user status, for instance location, mobility, and preference profiles. Reports about context-aware meeting systems adopt that interpretation at the expense of the context of the meeting itself.
- The majority of systems concentrate on what the technology can do for the user, rather than what the user can achieve. As an exception, Waibel et al, referring to the capabilities required for "interactive, integrated meeting support rooms" note with regret that "the technologies that provide such capabilities are as obstructive as they are useful – they force humans to focus on the tool rather than the task." [4]
- Adaptivity, logging, and trust management have all received some consideration, but not to the same extent as, for example, configuring smart environments.

The opening session of the first workshop turned out, in a sense, to be seminal. The facilitated discussion that followed the brief introductory presentations by each participant brought out four sub-themes as a basis for the successful planning and conduct of smart meetings:

- Joining up
- People
- Decisions and Provenance
- Capture and Retrieval

We explored the issues associated with those sub-themes and developed three key considerations that underpin the productive exploitation of smart spaces and smart technology:

- Designing
- Capturing and Analyzing
- Selecting and/or Exploiting

In our paper about exploiting smart meetings for knowledge transfer we examine the principal ideas associated with each sub-theme, and some of the questions arising from the three considerations [5]. Drawing on the experience of the two workshops we consider three aspects specifically related to knowledge transfer:

- Bringing the knowledge into the meeting space;
- Maximizing the benefits for the people in the space;
- Enabling people unable to be in the space to share the transferred knowledge.

Because we did not know at the outset of the first workshop the areas on which the participants would focus, we did not address remote participation specifically. We did however inform several people unable to be present about the Twitter feed, a small number of whom did use this means of making remote contributions. We are conscious that the calibre of human interaction will influence the effectiveness of knowledge transfer, as discussed in the third paper [1] and one of our goals for future work will be mechanisms for facilitating distributed knowledge transfer meetings, particularly in the context of sharing research outputs. The requirements for a *smart meeting log system* will include provisions for remote monitoring and contribution.

One issue that emerged from exploring the sub-themes was the meaning of the term *smart* and how we might distinguish a smart space from a 'dumb' space. The first workshop did not really tackle this issue, so we included it specifically in the agenda for the second workshop. A full discussion of the quality of *smartness* is beyond the scope of this paper, but in the following synopsis conveys our basis for regarding smartness as *conferred capability*.

No space is, or can be, inherently smart. Indeed the term could be regarded as an example of jargon that is acceptable because everyone thinks they know what it means. Without prejudice to any conclusions that we might draw in our planned review article, the overwhelming majority of the existing definitions express *smartness* in terms of technology and the capabilities it can confer: intelligence; assistance (to humans); and adaptivity (including mobility). A Smart Technology Research Centre [6] poster provided by Katarzyna Musial, one of the participants at the first workshop, lists five attributes of smart systems: adapting, sensing, inferring, learning, and anticipating. All five are associated with *key technologies*. The capability of a space ascribed as smart is infinitely variable, according to how that space is instrumented and configured. The *instruments* can be hardware or software, where the characteristics of the latter can range from passive service to intelligent agent; for a *system* to achieve smartness, we deem certain components to be essential, most notably people.

This view of smartness as *conferred capability* casts technology in a supportive rather than a controlling, or even mediating, role. What then of the role of the traditional capture methods, notes and flipcharts, and the use of a facilitator?

As well as the traditional methods, our methodology for capturing records of the workshop proceedings was based on recording technologies and social networking (Twitter). We hope in due course to evaluate fully the relative merits of all data capture methods used, both technological and traditional, but the following list comprises our provisional assessment of the key considerations:

- A comprehensive record of any meeting is arguably unattainable, given that individual video recording of every participant would be too intrusive.
- Audio recording alone misses the non-verbal communications that can sometimes be influential.
- A contemporaneous Twitter feed is beneficial, but can be distracting.
- Subsequently, the value of a meeting record depends upon a means of extracting points of interest efficiently and effectively. However, if the extraction process involves editing, the interpretation is likely to be influenced by the editor's perspective.
- Similar concerns arise with regard to the potential influence of the chairperson, the facilitator, and note-takers. Both workshops provided indicators of how the flow of a meeting might depend on such factors. Capturing a range of records, annotated with semantic links, is capable of providing an accessible and reliable resource.
- Emphasizing discussion over prepared presentations is beneficial. However, an explicit facilitator can influence both that discussion and the nature of the outputs, as indeed can the chairperson. It is clear that such roles include an editorial function, much like that of the editor of the final deliverables, but the influence of a facilitator can be much more subtle and less obvious.
- Any of the technologies we did or might have used has the potential to have an impact on the success of a meeting. For example, technology can assist the facilitator in enabling all participants to participate fully, which is a positive influence. On the other hand, technology can simply provide more routes for the loudest person to dominate and so reframe the discussions. Supporting technologies can remove barriers to participation but raise other new ones.

The rooms we used for the two workshops differed in several respects, such as: aspect ratio, openness, table layout, and – less tangibly – ambience. The Edinburgh room was square and spacious, whereas the Southampton room was smaller and more confined, particularly in its width. The facilitator (CB) was particularly conscious of the restriction, because it prevented him from engaging with all the participants at the same time. Such observations led us to recognise the potential significance of the physical space and the manner in which humans configure that space; humans who run meetings can exploit the characteristics of the physical space to influence both the conduct of meetings and their outcomes. We intend to explore this issue further when considering the human aspects of smart spaces [5].

4 Conclusions

With the two workshops we have experimented with space in the context of meetings that could be ascribed as smart, despite the lack of pervasive technologies. We learned lessons from these workshops that we intend to explore further to achieve improved and more dynamic techniques for recording the processes and outputs of research, particularly meetings about research.

With regard to the additional goals that we came to regard as meta-objectives, we believe it to be both necessary and appropriate to continue our investigations into the infrastructure required to support meetings in general, but particularly for research, and by extension learning environments. All three meta-objectives are highly relevant to such studies.

The two workshops bring out the point that the use of computers and technology in general are not ends in themselves. What we need is to find smarter ways of doing things that reduce the human effort and maximize the beneficial outcomes of meetings and discussions. We believe that the insights we gained from these workshops can influence strategies for exploiting smart meetings for knowledge transfer.

In the longer term, we believe that the current tacit acceptance that smart spaces somehow just happen can, and should, be replaced by an approach that relies on design and on empowering the users of smart spaces in that design. That, we hope will be the legacy of the Smart Spaces Theme, and not only for e-Science.

Acknowledgments. We acknowledge support for the Smart Spaces research theme from the e-Science Institute EPSRC EP/D056314/1. Our thanks also go to all the contributors to both workshops.

References

1. Frey, J., Bird, C., Willoughby, C.: Smart meeting spaces for knowledge transfer. In: InnovationKT 2012 (2012)
2. Research Themes, e-Science Institute, http://www.esi.ac.uk/research-themes (accessed November 15, 2011)
3. Smart Spaces for Smart People, http://wiki.esi.ac.uk/Smart_Spaces_for_Smart_People (accessed November 15, 2011)
4. Waibel, A., Schultz, T., Bett, M., et al.: SMaRT: The Smart Meeting Room Task at ISL. In: Proceedings ICASSP 2003 IEEE International Conference on Acoustics Speech and Signal Processing, vol. 4, pp. 752–755 (2003)
5. Frey, J., Bird, C., Willoughby, C.: Human aspects of smart spaces for knowledge transfer. In: InnovationKT 2012 (2012)
6. Smart Technology Research Centre, Bournemouth University, http://www.bournemouth.ac.uk/strc/ (accessed November 15, 2011)

Human Aspects of Smart Spaces for Knowledge Transfer

Jeremy Frey, Colin Bird, and Cerys Willoughby

Chemistry, University of Southampton, University Road, Southampton SO17 1BJ
{J.G.Frey,colinl.bird,Cerys.Willoughby}@soton.ac.uk

Abstract. During May and June 2011, we ran two workshops with a theme entitled "Smart Spaces for Smart People" [1]. Although organized under the auspices of the e-Science Institute, the participants came from a variety of disciplines and brought a range interests. The workshops themselves were run as experiments in running smart meetings with the intentions of exchanging and recording knowledge and decisions discussed in the meeting. A recurring theme in the workshops was not only that technology can be provided in a smart space to help in the knowledge transfer and recording process, but also that the technology will only be adopted and exploited if the users of the smart space can easily use it. There are other human factors that affect the success of collaboration in a smart space. These include the willingness for participates to collaborate if they have concerns over privacy and anonymity, particularly when discussions and decisions are recorded using technology. The dynamics of how participants work together in groups to transfer knowledge can also be enhanced through the use of smart spaces. The fact that the workshops were run in different physical environments also provided insights into how the physical design of the meeting space might have on effective collaboration and therefore effective transmission of knowledge. This paper is the second in a series of three, each dealing with different aspects of the workshops and how they influenced our thinking about knowledge transfer meetings, particularly in the context of sharing research outputs.

1 Introduction

A smart space is usually thought of as a meeting place where people come together to collaborate, to share knowledge, and engage in shared activities. The spaces are usually physical, for example, a meeting room, a classroom, a research lab, or museum; but they may also be virtual spaces, for example an online meeting environment. In a smart space the transfer of knowledge may have many different purposes, from the sharing of ideas and experience to solve problems or make decisions within a multidisciplinary organisation, the dissemination of knowledge in a research environment, to the packaging and transmission of knowledge for education to students or customers.

A search across the Internet or through the relevant literature for 'smart spaces' provides a myriad of definitions that equate smart spaces with the technology that is developed for them, for example ubiquitous computing, ambient intelligence,

R.J. Howlett et al. (Eds.): *Innovation through Knowledge Transfer 2012*, SIST 18, pp. 19–29.
DOI: 10.1007/978-3-642-34219-6_3

integrated devices and agents, and so on. The overwhelming focus on the technology can overshadow the human aspects of smart spaces that play an important role in the effectiveness and quality of knowledge transfer within such spaces.

Smart spaces and their associated smart technology can be used to facilitate the knowledge transfer process. However the effectiveness and quality of the knowledge transfer will depend upon the effectiveness of the smart space. Groups and individuals will only adopt smart things if they can easily exploit them and they add value to what they already do. Using the experience we acquired from running two *smart spaces* workshops, this paper explores some of the human aspects of smart spaces that succeed in helping users to collaborate and share knowledge, or alternatively, may inhibit their desire or ability to do so.

We ran the workshops under the auspices of the e-Science Institute with a theme title of "Smart Spaces for Smart People", with the original intention to explore interactions between the physical and digital worlds. The workshops also looked at strategies for successful planning and conduct of smart meetings by using the workshops themselves as an experiment in running smart meetings. The meeting deployed a variety of hardware and software in an attempt to capture the discussions in the meeting in various formats for later processing and use, for example, recording audio and video, or capturing notes and comments through different media including flipcharts used by the facilitators.

The workshop invited interested parties to attend and contribute their experiences and ideas about exploiting smart spaces and best practices. The discussion during both workshops was predominantly about the productive exploitation of spaces ascribed as smart, particularly the use of technology in making the space smart. However, a recurring theme from the workshops was the needs of the people using these spaces. Although we didn't set out to consider human aspects, the experience of running the workshops as smart meetings in locations with different technological challenges, provided an opportunity to observe the relative smartness or dumbness of these particular spaces and their facilities extending beyond just the functional capabilities of the supporting technology.

2 The Pain of Technology

Running a meeting where all the participants are in the same room should be a very simple exercise, but as demonstrated in the workshops can in fact become very complicated because of the difficulty of using technology. Successfully connecting a laptop to a projector to display some slides can often result in a small delegation of hopeful participants poking at buttons, pulling and pushing wires whilst the hapless presenter presses key combinations and maybe even reboots their machine. The speaker or other users may desire network or Internet connectivity that may take time and assistance to configure. All this activity wastes times, inevitably causes the speaker some embarrassment, and is disruptive and distracting to the purpose of the meeting. The technological frustrations are magnified with more complex gatherings such as teleconferences, videoconferences and web meetings when meeting participants are located in more than one location. There are features of the technology that would assist in knowledge transfer within the meeting, or for participants that were unable to take part at the time, but that are never utilised because of the complexity involved.

In the workshops we certainly experienced some of the pain of technology that could be described as contributing to the dumbness of the space, rather than the smartness of it. In the first workshop the main technical difficulties revolved around making the audio and video equipment work. There were many more problems in the second workshop around more standard technology. For example, because many of the attendees were not present at the previous workshop, it was essential to share the background and experiences of the first meeting. The materials had been prepared and presented from a website, so the meeting couldn't start until everyone had wireless access, but this required device by device approval using mac addresses. This was complicated, time-consuming, relied on a single person to do it. These challenges with the technology resulted in disruption, distraction, and delay. All factors contributing to a potential failure of the planned knowledge transfer because of distraction and time limitations.

If these are problems experienced with familiar technology in a 'normal' meeting room, what about really 'smart' spaces? Similar problems affect smart high-end room systems that often have multiple displays, interactive whiteboards, robotic cameras and remote conferencing systems. Research shows that these technologies generate two main problems [2]. The first that users fail to engage with the technology, they don't know what technologies will be in the room, let alone how to exploit them. The second problem is that the technology in these rooms is so complex they need a resident expert or 'wizard' to maintain the technology, and to help the users to use it. These wizards have to be around for the room to be useful, again resulting in a loss of potential opportunities and effectiveness of knowledge transfer and collaboration.

2.1 Change the User or the Technology?

One suggestion from the workshops for overcoming the problem of hard technology is to train the users in the 'new language' in order to make it easier to design and create technology. This idea suggests that people's behaviour needs to be changed in order to make it easier for the technology! Should we really expect user behaviour to change so that the technology can understand the user? Although undoubtedly people's behaviour does change in reaction to technology, the change is difficult to predict at best, and almost certainly impossible to control. A user's frustration comes from needing to change the way they behave to fit the technology. The users need to understand how it works in order to use it. Although users can learn new interactions, those that are difficult or unnatural are more likely to lead to avoidance behaviour in the users. The goal of the user is also important in this discussion, for a role such as trainer or teacher, spending the time learning to use the technology to facilitate knowledge transmission makes a lot of sense, but in other contexts spending time learning the technology may have no personal value. As an example, participants have no personal benefit in learning to use audio or visual recording equipment for a meeting to support knowledge transfer for participants not present at the time.

2.2 Fitting into the Human Environment

A different suggestion for solving the problem of using technology in smart spaces is the progression of technologies that recognize human actions for the development of 'natural interfaces'. These technologies include motion tracking, gesture recognition, face expression, and gaze-aware interfaces, in addition to taking input from the users using more naturalistic methods such as captured writing and speech recognition [3]. These technologies allow the users to actively use actions and gestures that they are familiar with, or that passively observe user behaviour and change the environment based upon it. In addition to providing technology that can support the users in their collaborative or knowledge transfer activities, these technologies may be able to non-invasively capture information such as reactions and user behaviour that could be used to assess the effectiveness of the knowledge transfer or collaborative activities occurring within the smart space.

3 Collaborating in a Smart Space

There are ways that a smart space can help users to collaborate who might find contributing in an ordinary meeting situation difficult. However, there are also aspects of smart spaces that may inhibit contribution by users.

3.1 Privacy Concerns

A recurring theme that came out in the workshops was privacy, particularly as a consequence of the recording or monitoring of participants in a smart meeting. Key concerns regarded whether the data would be made publically available, who had access and control of the data, how was it going to be used, and what happens to the data once it has been finished with? In the workshops it was made very clear that audio and video recording was taking place. Participants agreed because they knew it was a part of the workshop, but also because the recorded data would remain private.

Initially the recording technology was very prominent, but by the second day the participants agreed that they were used to it and ignored it. The technology faded into the background and became a part of the space.

People often make their own notes on the conversation or the decisions that are made in a collaborative situation or meeting. In some situations, usually more formal meetings, a note taker is present to formally record what was said. These note-takers are in effect performing the same task as an audio or video recording, although potentially less accurately. Despite the better reliability of the recording taken with technology, the members of the workshop felt more comfortable and less concerned about their privacy with the recordings taken by the human note-taker rather than with the technology. Another advantage of the written record is that is it more accessible in the event of a query. Locating and replaying the segment of an audio record is difficult, especially if is has not yet been annotated, which is itself a difficult task.

3.2 Use of Twitter

Users in a smart space may have access to an Internet connection through a laptop or even via mobile devices. People can easily communicate their activities and words that have been said in that space to others who are not present. Twitter enables people to broadcast to anyone in the world what they are doing. There are examples where 'tweeting' of details of discussions or commentaries on presentations have had both positive and negative effects.

Twitter use was actively encouraged during the workshops and the results retained as part of the meeting record. Tweeting can be seen as an effective way to elicit interest, feedback, knowledge and experiences from the wider community when the tweets contain the right kinds of questions, links, or status remarks. The use of social networking in general within a meeting enables the participants to extend their knowledge beyond their own experiences and tap into the knowledge of the wider community. These interactions and sharing can extend beyond the boundaries of the meeting in both participation and time.

In the workshops, tweeting from within the workshop encouraged participation from people not in the room, including some who had never even heard of it before.

The same privacy concerns that are expressed for recording or monitoring activities in a smart space also exist for technologies such as Twitter, because the perceived privacy of the collaboration for one user may not match the expectations of another who can instantly share their thoughts about the collaboration with the rest of the world. Through the use of 'retweets' a tweet can spread virally in a matter of minutes with no way to stop it once it has started.

3.3 Anonymity

There is a potential conflict in any knowledge transfer environment if a participant desires anonymity. The use of smart technology has the potential to increase this conflict. A smart space can be much more smart if it can recognize and 'know' information about the individuals within that space, for example the smart space can support the users by providing context based on the needs of the individuals within the group. However, collaborators may want to share knowledge, experience or opinions, but feel more comfortable doing so anonymously. Often though the capture and dissemination of knowledge, ideas, and opinions does not requiring a need to know who gave the information. Not knowing the identity of the knowledge provider may mean the loss of certain context, but at the same time providing the ability to contribute anonymously can facilitate communication and honesty.

Software systems that allow the capture of information anonymously have been available and used in focus groups situations for decades [4]. These have an obvious benefit in group collaboration situations, but they are not yet in general use in smart environments.

3.4 Improving Collaboration Using Smart Spaces

The effectiveness of collaboration in a group is in part determined by the composition of the group, how equal the members are, and how much they have in common. Groups can be classified according to the following characteristics of their membership [5]:

- Homogeneous groups, where members are equals and have the same privileges
- Heterogeneous groups are unequal and members have different privileges
- Loosely coupled groups
- Tightly coupled groups

Knowledge transfer may involve any of these combinations of groups. For example, attendees at a conference are usually homogeneous but loosely coupled. They are equal as peers, but don't necessarily know many of the other attendees. A research team on the other hand may be homogeneous, but also tightly coupled. Homogeneous groups tend to have similar experiences and knowledge, and understand situations and facts within the same contextual framework potentially making knowledge transfer simpler.

Within organisations groups are typically heterogeneous, for example in multidisciplinary teams the team members have different roles with different levels of influence and authority. If they work closely together they will be tightly coupled, but could be loosely coupled if they have been brought together to solve a problem from different parts of an organisation. Knowledge transfer is likely to be easier in closely coupled teams even if they are multidisciplinary because they are likely to share the same goals and same context for the knowledge transfer.

In a learning environment the group is also be heterogeneous, the teacher has more authority than the learners. The group may or may not know each other. A school class may be tightly coupled compared to an adult education college where the learners are only loosely coupled. In a learning environment the knowledge transfer is somewhat one sided and likely to be limited in scope, but is very effective for the transfer from teacher to learner because the roles involved and the context are very well understood.

Another example of heterogeneous and loosely coupled groups may be multicultural groups. In a meeting environment the participants may have different abilities in the language being used to conduct the meeting. There may also be cultural differences such as a need for more or less context in the communication. This can lead to disadvantages for some in following the conversation, being able to contribute to the activity, and even potential misunderstandings.

Multidisciplinary teams may struggle to collaborate or share knowledge because unfamiliar jargon or terminology with different meanings may cause communication difficulties and misunderstandings. The context that the people in these roles have may be very different. It is not effective to simply share the information that is familiar and meaningful for your own role in order to help someone in another role to understand it. In heterogeneous groups problems may arise where the perceived knowledge or authority of some individuals are greater than another. Those perceived as having the greater knowledge or authority may be listened to more, whereas others with equally valid ideas or experiences may not have the chance to share them, or may be held in lower regard. The less knowledgeable or lower ranking participants may feel inhibited and unwilling to share their knowledge because of this effect.

In loosely coupled groups communication difficulties or inhibitions can result from the members having a lack of knowledge about the other members of the group such as their background or shared interests. Others who may find it difficult to contribute are individuals with a disability, who are shy, or whose first language is not the same the language that the activity is being conducted in.

Knowledge transfer in large organisations where there may be many ranks or authority, levels of expertise, and distributed team members may struggle with knowledge transfer activities because of these types of interactions within groups. The groups are often multidisciplinary, hierarchically organised, and come from different backgrounds, cultures, and may even speak different languages. These problems can be seen in knowledge transfer between an organisation and its customers because their needs and level of understanding are so different.

A smart space can help solve some of these common problems of collaboration by ensuring the group and individuals have access to the information they need (and only the information they need) in the formats that are most appropriate to them, based on language, abilities, and role. Knowledge can be presented to the user in the form they are used to working with, for example, blueprints for an architect and MRI scans for a surgeon. The knowledge can also be made available in accessible formats such as in the user's mother tongue language, or in large print or audio. The smart space must have an awareness of the users in the room and their needs, as well as access to the relevant information and the different ways of presenting it.

For distributed multidisciplinary groups the use of virtual smart spaces can be invaluable as ways of enabling interaction between the members of the groups and the transmission of knowledge across an organisation. Important technology that is becoming more widely used in such virtual smart spaces includes real-time translation services, access to background information, as well as the usual methods of sharing information, and recording the visual, audio, text, and presentation elements of a meeting. These technologies can improve the accessibility and understanding for all the participants regardless of location and background.

The recording of activities and discussions in a virtual space is valuable for participants, for example those who struggle to keep up with the conversation in a meeting, or for those participants in different time zones, who may be unable to attend the meeting in person, but who need to acquire the knowledge from the meeting. Virtual smart spaces can have an existence that outlives the time of the meeting that actually occurred, enabling participants to continue their conversations, thoughts, and ideas over a period of time, and facilitate knowledge transfer to new comers on a project.

The workshops intentionally involved face-to-face participation, and there were no remote participants, but there were discussions about the difficulties of distributed meetings. Those meetings that are supported by smarter technology are more successful and lead to a better quality of knowledge transfer than those where limited technology is available, for example web conferences with audio are more successful for meetings and education than audio conferences alone.

Within the workshops the teams were heterogeneous in some respects being from different backgrounds and disciplines, but there was a lot of commonality in terms of knowledge, goals, and interest in the topics of the discussion. The experiences of participants discussed in the workshops emphasised the difficulties in working with heterogeneous teams. One of the participants from the first workshop raised the kinds of issues often seen in heterogeneous teams, when describing their experiences working with multidisciplinary teams in a medical context. Specifically that it is a struggle to work in a multidisciplinary team, and in teams where everyone wants different information.

In the workshops there were differences between levels of participant interactions and also the methods of interaction. For example, some participants tended to verbalise their thoughts and ideas, whilst others wrote their own notes or made use of technology to capture and express their thoughts, opinions, and ideas, for example Twitter. Some individuals shared more knowledge than others. Although many of the reasons why some participants interacted more than others can be down to the dynamics of group behaviour, there are some elements of the space themselves that influenced the behaviour of the participants and the effectiveness of knowledge transfer in the workshops.

4 Physical Characteristics of Smart Spaces

An unexpected aspect of the workshops was the discovery that the physical characteristics of a space may influence the success of group collaboration. These physical characteristics can contribute to the 'smartness' or 'dumbness' of a space. The difference in the facilities between the two workshops highlighted physical characteristics that impacted on the effectiveness of the venues for collaboration including lighting, space, layout, and noise. These physical characteristics of a space affect the behaviour of the people within the space and therefore have an impact on the effectiveness of collaborative and knowledge transfer activities that take place within the space. Consider, for example, how the arrangement of chairs at an interview, the relative height of the chairs, relative position, and so on, can affect the relationship between the interviewer and interviewee. Chairs laid in a circle are more likely to encourage people to express their opinions than a forma layout. An individual can be made to feel more or less powerful, or more or less in conflict with another person in the room by something as seemingly trivial as the layout of chairs. The impacts of other physical factors are more obvious, for example, if the room is poorly lit or there is a loud background noise, individuals with visual or audial problems may find it difficult to follow a discussion or presentation. There is also potential for the physical layout of the space to be exploited by individuals for their own ends, which may lead to both positive and negative consequences.

The room in for the first workshop at Edinburgh was very spacious with room for everyone at the table. Layout of the room enabled the flipchart to be clearly visible at the front. This provided a central focus, and people weren't distracted.

The space was also extended because the participants could leave the room for breaks and lunch. This extension of space actually facilitated constructive discussions because the conversations were continued into the breaks (although many of these discussions and their ideas were probably lost with no mechanism for recording). Physical problems at Edinburgh were to do with the organisation of the tables. The tables were organised over the power supply hatches on the floor, so the tables had to be moved before the beginning of the meeting. Some of the chairs were a bit constrained by the legs of the tables. Square tables also made it difficult to see everyone and see who was talking. Overall though, the room facilitated effective knowledge transfer and recording.

The second workshop at Southampton, in contrast, was not a good example of a smart space. There were technological problems as previously discussed, but there were also logistical problems based strongly on the physical characteristics of the room. The room was very small and cramped, the layout was poor, and there were not enough seats for all the participants at the table. There was very little space even at the table, and you could not see the other people around the table, only the person next to or opposite you. Those people who were not sat at the table were effectively removed from the discussion. Other physical characteristics that contributed to the confined feeling of the room included poor lighting and noisy air-conditioning. There was no extension of that space, because all the activities were carried out in the room including breaks and lunch.

Furniture layout at Southampton also contributed to a colocation of activities, the flipchart was in the corner, and was therefore very difficult to see. There were also different things going on at the front, with two screens, one with a twitter feed and the other with the agenda or presentation plus the discussion at the tables. The participants of the meeting found the twitter feed distracting because of the movement, and the difficulty of reading something different to what is being said at the time. The multiple locations of activity resulted in multiple focuses of attention. This contrasts with the first workshop where the location of the flipcharts and facilitator was more successful, providing a stronger and more visible focus of attention.

It is difficult to make a direct comparison between the two meetings to know whether the physical conditions made a significant difference to the way that the meeting worked or the outcomes. Various differences between participant behaviour and contribution in the workshops could have been affected by these physical differences. For example, because not all the participants were sat at the table in Southampton, then not everyone was equally able to share their knowledge and contribute to the discussion. The impact of physical characteristics of a smart space on the success of a meeting would make an interesting topic for investigation in the future.

If a smart space and related technology are to be used effectively by a particular audience, then the needs and desires of that audience must be understood. If this is not done, users will avoid using the parts of the space that are difficult or that they do not see the need for. This can result in avoidance of the entire space. Ideally there should be participatory design where the future users of a space are involved in the design. The design of smart spaces will be better and more likely to be a success if the designers experience the way that the users work now, and embedded themselves in that culture and environment.

5 Conclusions

A smart space is only smart if it enables the users of that space to use the space for its intended purpose. Participants must be able to effectively collaborate, share, and engage in knowledge transfer activities. Technology in smart spaces can be used in numerous ways to support participants and enhance the quality of knowledge transfer. There are many tools that can help facilitate the knowledge transfer by providing greater accessibility to the knowledge and presenting the relevant knowledge in the most useful context for each individual, as well as the group. Technology can be used to assist those who have a disadvantage in a group, for example those with language difficulties, disabilities, or less experience within the group.

The kinds of groups that are most likely to benefit from the support and assistance are also those heterogeneous groups that are also the most likely to be concerned about the privacy implications of using it. I am more likely to trust the other participants in the smart space if I already know them or we have a lot in common. Considerations need to be given to the potential privacy issues and how a particular group may react. It may be appropriate to enable participants to contribute anonymously if they are more likely to share their knowledge, opinions, and experiences under those conditions.

Although technology can be of great benefit to knowledge transfer within a smart space, it is important to remember that technology that is hard to use or understand is at risk of not being used correctly or even not used at all. Any benefits that technology may have had will be wasted, and can even have a negative impact on knowledge transfer by distracting the participants or taking valuable time away from the key activities through technical difficulties. Smart technology is technology designed with an understanding of the goals and needs of the users, and ideally with user involvement that fits into the human environment.

However, it takes more than well-designed technology to make a space smart. Even with the most functional, easy to use, and relevant technology in the smart space does not guarantee successful collaboration. Something as simple as the position of chairs around a table can enhance or stifle the inclusion of participants, and consequently the success of the knowledge transfer that occurs.

In conclusion, the physical characteristics of the space together with the usability and 'calmness' of the technology contributes significantly to the relative smartness or dumbness of the space.

Acknowledgments. We acknowledge support for the Smart Spaces research theme from the e-Science Institute EPSRC EP/D056314/1. Our thanks also go to all the contributors to both workshops.

References

1. Frey, J., Bird, C., Willoughby, C.: Smart meetings: experimenting with space. In: InnovationKT 2012 (2012)
2. Back, M., Golovchinsky, G., Boreczky, J., Qvarfordt, P., Denoue, L., Dunnigan, T.: The USE Project: designing smart spaces for real people (2006), `http://www.fxpal.com/UbiComp2006/USE_ubicomp.pdf` (retrieved July 2011)
3. Cook, D.J., Das, S.K.: How smart are our environments? An updated look at the state of the art. Pervasive and Mobile Computing 3, 53–73 (2007)
4. Group Systems software, `http://www.groupsystems.com/`
5. Wang, Z., Zhou, X., Yu, Z., Wang, H., Ni, H.: Quantitative Evaluation Of Group User Experience In Smart Spaces. Cybernetics and Systems 41(2), 105–112 (2010)

Smart Meeting Spaces for Knowledge Transfer

Jeremy Frey, Colin Bird, and Cerys Willoughby

Chemistry, University of Southampton, University Road, Southampton SO17 1BJ
{J.G.Frey,colinl.bird,Cerys.Willoughby}@soton.ac.uk

Abstract. During May and June 2011, we ran two workshops with a theme entitled "Smart Spaces for Smart People". Although organized under the auspices of the e-Science Institute, the participants came from a variety of disciplines and brought a range interests. We placed a strong emphasis on facilitated discussion, with the clear intention to explore ideas about exploiting the interactions that could occur within smart spaces. Although the workshops formulated the view that no space is, or can be, inherently smart, we deemed certain components to be essential for a system to achieve smartness, most notably people; the role of hardware and software technologies is to confer capability. The lessons we learned are applicable to any smart meeting. We grouped our findings under four sub-themes that we identified as a basis for the successful planning and conduct of smart meetings. After examining the principal ideas associated with each sub-theme, we go on to consider how these ideas might influence strategies for exploiting smart meetings for knowledge transfer. This paper is the third in a series of three, each dealing with different aspects of the workshops and how they influenced our thinking about knowledge transfer meetings, particularly in the context of sharing research outputs.

1 Introduction

Knowledge transfer is a practical problem to which solutions can involve a range of methods and activities. In this paper, we focus specifically on meetings as activities contributing to the knowledge transfer process. Using the experience we acquired from running two *smart space* workshops we aim to explore how conducting meetings in smart spaces might enhance the knowledge transfer process. We ran the workshops under the auspices of the e-Science Institute with a theme title of "Smart Spaces for Smart People", with the original intention to explore interactions between the physical and digital worlds. However, the emphasis shifted to the productive exploitation of spaces, especially meeting spaces, ascribed as *smart*. We describe the workshop methodology in the first paper of the series [7] and explain how we collated the results and assessed the outcomes of the two workshops. In the same paper we clarify our basis for regarding smartness as *conferred capability*.

Although the aims and objectives of the workshops expanded, our initial approach was to consider smart environments in general, but with specific attention to research, teaching, and meeting spaces. For both workshops, we invited

R.J. Howlett et al. (Eds.): *Innovation through Knowledge Transfer 2012*, SIST 18, pp. 31–38.
DOI: 10.1007/978-3-642-34219-6_4

participants with a range of interests, some of whom attended both events. We placed a strong emphasis on facilitated discussion, apart from the opening session, which included brief introductory presentations. The meeting rooms for both workshops contained audio-recording devices and for the first workshop, we also used video recording. Both meeting rooms offered wireless access to Twitter; for the second workshop, we also projected the Twitter feed onto a screen. Our use of Twitter had several underlying purposes, all of which are capable of being relevant to knowledge transfer:

- We wanted meeting participants to be able to capture thoughts and ideas at the time they occurred, but without interrupting the flow of the discussion.
- We wanted a time-stamped trace of the key points in the discussion, especially the perceived turning points.
- We wanted external parties who were interested in the workshops but unable to be present to be able to inject observations and questions.

Tweets from remote followers were few in number, but were usually relevant responses to points made in tweets from the meeting participants. Such use of Twitter could be valuable for knowledge transfer sessions. However, in our subsequent analysis of the workshops, we were aware that a contemporaneous Twitter feed could be distracting as well as beneficial.

For both workshops, we also used an independent note-taker, and the same person (CB) acted as facilitator for both workshops, using flipcharts to capture the key points raised during proceedings. These traditional capture methods were not only complementary to the audio- and video-recordings but also provided valuable insights into the factors important for effective meetings. The evaluation sessions of both workshops concluded that, in broad terms, having a facilitator was valuable for the conduct of what were intended to be *smart* meetings.

An important motivation for organizing the workshops was to explore techniques to improve the recording of research processes and outputs. We see a helpful analogy between knowledge transfer and the sharing of research outputs, allowing that the two endeavours might be on different scales.

We recognise that neither meetings in general nor the use of smart meeting spaces offer solutions for all aspects of the knowledge transfer problem. To cite just one example, it is unlikely that any smart system will materially assist the elicitation of tacit knowledge. Indeed, we have an intuitive concern that some individuals, when in a space ascribed as *smart*, might feel inhibited about disclosing some or all of their tacit knowledge.

2 The Role of Meetings in Knowledge Transfer

We fully accept that meetings are but one of a range of activities associated with knowledge transfer: the process is broad in scope and organizations have devised several ways to effect knowledge transfer. The transfer process involves more than just communication, but the effectiveness of the latter can materially affect the perceived satisfaction with the process. For the purposes of this paper, we assume that the context of transfer is cooperative and that the structure and organization of the knowledge itself do not present hurdles to the process.

Knowledge transfer is more than a form of training, although learning is a key part of the process for the recipients. If the medium for knowledge transfer is a meeting, it is important to create a supportive learning environment. The University of Southampton encourages students to design smart learning spaces for themselves with the "Create Your Campus" competition [1].

Clearly, the calibre of the interactions between participants determines the overall effectiveness of any meeting. Reviewing the knowledge sharing literature in 2003, Cummings examines a variety of factors affecting knowledge sharing [2]. Considering physical distance, he explains the evidence in favour of face-to-face meetings facilitating knowledge transfer in terms of the relationships between the parties.

The effectiveness of knowledge transfer meetings will also depend on the support provided. The UbiMeet workshop [3,4] explored this issue, noting in the overview that, for meetings to be more engaging "requires relaxing the notion of meeting support from a particular panoply of conferencing and annotation technologies to more broadly any set of tools that enable synchronous communication amongst a group as well as tools that can help compensate for differences between different people's situations."

Unsurprisingly, the emphasis of UbiMeet was on technology, and solutions based on ubiquitous computing. Traditional forms of support, involving humans, also have a significant role, such as that of a *facilitator*. While numerous sources consider the role of a knowledge sharing facilitator, the key lessons are presented in a USAID wiki, derived from a program set up to assist agencies in Building more Effective Learning Organizations, the BELO program [5].

Knowledge transfer meetings can be non-formal, as exemplified by the Open Space method [6], which encourages groups to define their own tasks and to adopt their own approach to dealing with those tasks. The facilitator has a key function in bringing the group together, identifying the task, and steering the group towards completing the task.

3 The Conduct of *Smart Meetings*

For both workshops, it was always our intention to explore ideas about exploiting the interactions that could occur within smart spaces. In our paper about experimenting with smart meeting spaces, we expound the view that no space is, or can be, inherently smart [7]. We deem certain components to be essential for a system to achieve smartness, most notably people; the role of hardware and software technologies is to confer capability.

The lessons we learned are applicable to any smart meeting, regardless of its purpose. We identified four sub-themes as a basis for the successful planning and conduct of smart meetings:

- Joining up
- People

- Decisions and Provenance
- Capture and Retrieval

We also developed three key considerations relevant to those sub-themes:

- Designing – applies to all four sub-themes
- Capturing and Analyzing – applies particularly to *Decisions* and *Capture*
- Selecting and/or Exploiting – applies particularly to *Joining up* and *People*

We now examine the principal ideas associated with each sub-theme, and some of the questions arising from the three considerations. In the next section we will go on to consider how these ideas might influence strategies for exploiting smart meetings for knowledge transfer.

Joining up means ensuring that the processes of the meeting are as seamless as possible and minimally intrusive: participants should be free to focus on the project rather than the process, a goal equally appropriate for the *People* sub-theme. A joined-up meeting uses the collective power of the participants, local and remote, and including where applicable their social networks. To assist joining up, any applications should be collaborative and interactive, and any technologies deployed should be interoperable.

From the *People* perspective, the goal must be to facilitate human-human interaction and communication, which can be assisted by using calm (non-intrusive) technology. It will also be important to ensure accessibility and to promote reward and recognition for the contributions made by the meeting participants. We explore these ideas further in our paper about the human aspects of smart spaces [8].

An implicit function for the smart meeting support is to capture the cycle of information, which involves recording context, evidence, and identification. In most, but not all, cases, it will be important to monitor participation: who did and said what. These functions underpin the *Decisions and Provenance* sub-theme.

The *Capture and Retrieval* sub-theme is concerned with the more tangible products of the smart meeting. Data, information, and knowledge are all important, but it is the *metadata* that is likely to be most valuable for the subsequent retrieval, reuse, and repurposing of the knowledge transferred. Capturing a range of records, annotated with semantic links, is capable of providing an accessible and reliable resource.

Although we have noted the importance of using non-intrusive methods, 'constructive mediation' has a vital role nevertheless. Quoting from one of the lessons of the BELO program [5], "facilitators are crucial to engaging staff and keeping momentum."

The questions arising from the three considerations were numerous. We reproduce here a small selection, chosen for their potential relevance to exploiting smart meeting spaces for knowledge transfer:

Designing:

- When is a space *good enough* (to enable progress)?
- How do the space and activities interact? Is customizing appropriate?

Capturing and Analyzing:

- What is the context and what are the purposes of capturing?
- To what extent are unanticipated uses catered for?

Selecting and/or Exploiting

- What should be the balance between selecting resources for addition to a space and exploiting the characteristics of existing spaces?
- Do spaces become smart – or smarter - as people enter?

4 Strategies for Exploiting Smart Meetings for Knowledge Transfer

It is debatable how often the ideas discussed in the preceding section are realized in practice, but they are nevertheless capable of informing discussions about strategy. We consider three aspects:

- Bringing the knowledge into the meeting space;
- Maximizing the benefits for the people in the space;
- Enabling people unable to be in the space to share the transferred knowledge.

Simplistically, we can refer to these aspects as strategies for before, during, and after the meeting. An important principle is that the meeting space exists, notionally at least, throughout the transfer process, and not just while the meeting itself is in progress.

Applying that principle to bringing the knowledge into the space, prior preparation, capture of data and metadata during the meeting, and analysis after the event are all parts of the process. We describe these activities as *continuous curation.*

Here we are using the term *data* in a generic sense, allowing it also to encompass information and even knowledge. Awad and Ghaziri place these concepts in a pyramid, describing *information* as data "in formation" and *knowledge* as actionable information [9], as shown in Figure 1. Bellinger, Castro, and Mills propose *connectedness* as the key to understanding meaning [10], although we prefer to label the vertical axis in Figure 1 explicitly as *understanding*.

For a joined-up knowledge transfer process, we need access to the data and information that lies below the knowledge in the pyramid, possibly in more than one form. We need metadata to facilitate retrieval, and cross-reference links to enable exploration. To support continuous curation we need a *meeting log system*, a tool that we were very conscious of lacking during both workshops.

Such a log system would combine meeting minutes with short contributions from meeting participants, keeping entries in chronological order but allowing links to previous entries, and enabling tagging with keywords and filtering of entries by author. A log system would also provide the natural foundation for the comprehensive record, with semantically rich links to the data captured by the other technologies. We are not aware of any tool currently in existence that meets all the requirements we envisage for a meeting log system.

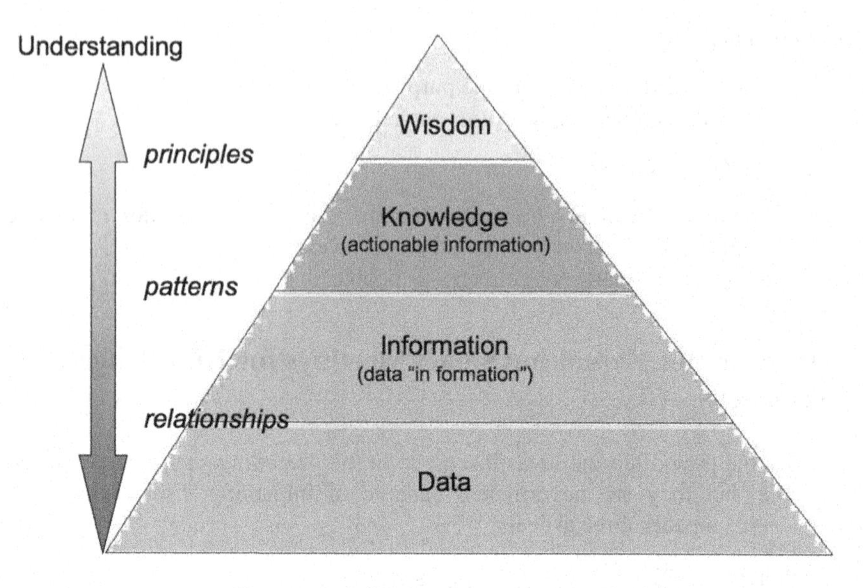

Fig. 1 The Data, Information, Knowledge, Wisdom (DIKW) pyramid

Linking is particularly important after the meeting, and even more so when transferring knowledge. Analysis after the event almost inevitably introduces editorial influence. Cross-links protect against consequential misunderstandings, because they allow people to check the edited record against the raw data.

People present in the space – participants in the knowledge transfer meeting – have an advantage with regard to confirming accuracy and, if necessary, veracity. People who were unable to be present but still require the knowledge will rely on links for navigating the meeting record. For example, a person who is unclear about the meaning of a particular item can trace back to the audio-record to listen to what the speaker actually said about the item. The person can also refer to the meeting log for any short contributions made during the period for which the item was under consideration. Moreover, people inspecting the knowledge transfer record after the meeting should be able to annotate the record. By doing so, they further enrich the knowledge: continuous curation never stops.

Although the workshops were not run on a large scale, with participants in several, distributed, locations, we remain conscious of the need to ensure that all contributions, however made, are effective, and that the record of knowledge transfer is accessible to people who were unable to be present but need to refer to that record subsequently.

5 Conclusions

We have analysed the results of two workshops that we ran with the intention to explore ideas about exploiting the interactions that could occur within smart spaces. Building on those results, we have presented an assembly of ideas associated with the effective conduct of smart meetings, presenting these ideas under four sub-themes that we identified as a basis for the successful planning and conduct of smart meetings. We have also posed a small selection of questions that arise from three key considerations that the workshops found to be relevant to those sub-themes.

We have proposed a set of strategies for exploiting smart meetings for knowledge transfer, derived from our analysis of the workshop results and the ides that emerged. Although we base our strategies on ideas explored rather than experiments conducted, nevertheless we consider that the suggestions embodied in those strategies do offer routes to enhancing the knowledge transfer experience and improving the satisfaction of the participants in the transfer process, before, during, and after the smart meeting.

Acknowledgments. We acknowledge support for the Smart Spaces research theme from the e-Science Institute EPSRC EP/D056314/1. Our thanks also go to all the contributors to both workshops.

References

1. Create Your Campus, http://www.southampton.ac.uk/music/news/2011/10/31_touching_sound_at_create_your_campus_launch.page? (accessed November 23, 2011)
2. Cummings, J.: Knowledge Sharing: A Review of the Literature (2003), http://lnweb18.worldbank.org/oed/oeddoclib.nsf/DocUNIDViewForJavaSearch/D9E389E7414BE9DE85256DC600572CA0$file/knowledge_eval_literature_review.pdf (accessed November 23, 2011)
3. UbiMeet @ UbiComp 2007 Workshop on Embodied Meeting Support: Mobile, Tangible, Senseable Interaction in Smart Environments (2007), http://www.fxpal.com/UbiComp2007/ (accessed November 23, 2011)
4. Back, M., Lahlou, S., Carter, S., et al.: Embodied Meeting Support: Mobile, Tangible, Senseable Interaction in Smart Environments. In: UbiMeet Workshop at UbiComp (2007), http://www.fxpal.com/publications/FXPAL-PR-07-408.pdf (accessed November 23, 2011)
5. Role of a Knowledge Sharing and Learning Facilitator. USAID wiki, http://apps.develebridge.net/amap/index.php/Role_of_a_Knowledge_Sharing_and_Learning_Facilitator (accessed November 23, 2011)
6. Open Space Technology, http://www.openspaceworld.org/ (accessed November 23, 2011)
7. Frey, J., Bird, C., Willoughby, C.: Smart meetings: experimenting with space. In: InnovationKT 2012 (2012)

8. Frey, J., Bird, C., Willoughby, C.: Human aspects of smart spaces for knowledge transfer. In: InnovationKT 2012 (2012)
9. Awad, E.M., Ghaziri, H.M.: Knowledge Management. Pearson Educational International, Upper Saddle City (2004)
10. Bellinger, G., Castro, D., Mills, A.: Data, Information, Knowledge, and Wisdom (2004), http://www.systems-thinking.org/dikw/dikw.html (accessed December 1, 2011)

Facilitating Knowledge Transfer in IANES - A Transactive Memory Approach

Matthias Neubauer, Stefan Oppl, Christian Stary, and Georg Weichhart

Department of Business Information Systems – Communications Engineering
Johannes Kepler University of Linz, Austria

Abstract. 'Interactive Acquisition, Negotiation and Enactment of Subject-oriented Business Process Knowledge' is a 4-year research effort to implement a knowledge life cycle using Subject-oriented Business Process Management and mutually align related Organizational Learning techniques in respective development processes. As different partners from academia and industry need to share their experiences, tools, techniques on a detailed level as well as the project's content management are of crucial importance. In this paper, we focus on the Nymphaea system as a means for effective spatially distributed knowledge sharing. We develop the requirements revisiting distributed transactive memory systems and describe the toolset and its support for different aspects of knowledge sharing within the project. We also report on initial findings when utilizing annotation features for individualization and mutually changing perspectives.

1 Introduction

In joint research projects between universities and industry, such as IANES[1], work is organized in a spatially distributed setting using several means to facilitate knowledge transfer among the partners. The main enabler of knowledge transfer is the exchange of research staff among the partners, enabling mutual learning processes in the course of the actual research and development. Knowledge transfer workshops are additionally conducted to create a basic common understanding about the current state of work and future tasks. They supplement the 1:n setting in staff exchange measures and extend learning processes to a wider audience. Knowledge transfer in both settings relies on the concept of *transactive memory* [13]. Transactive memory is a conceptual type of memory (i.e. stored knowledge) besides the (individual) internal memories and (codified) external memories and refers to *"a set of individual memory systems in combination with the communication that takes place between individuals"* ([13], 1987, p. 186). Knowledge transfer here is bound to the opportunity of direct interaction among the involved people.

---

[1] "Interactive Acquisition, Negotiation and Enactment of Subject-oriented Business Process Knowledge", funded under the FP7-IAPP (Industry-Academic Partnerships and Pathways) programme.

R.J. Howlett et al. (Eds.): *Innovation through Knowledge Transfer 2012*, SIST 18, pp. 39–50.
DOI: 10.1007/978-3-642-34219-6_5

Consequently, the organizational coverage of the two measures described above to facilitate knowledge transfer is limited in spatially distributed work settings. Operative work processes are organized in an asynchronous, distributed way. Information made available by partners has to be accessible in these work situations even in situations, where no person with the necessary skills or knowledge is available on-site. Furthermore, certain groups of people can hardly be reached with traditional knowledge transfer measures targeting towards academics. Especially at industry partners, people without an academic background but a vast amount of practical experience maypossess highly relevant knowledge and, as such, their involvement can be crucial for project success. In IT-focused projects, e.g. non-academic developers are affected. They have to understand the project requirements provided by research and are also sources of knowledge, regarding, for instance, technical architectures or interfaces. In order to cover distributed asynchronous knowledge sharing scenarios and involve otherwise excluded people in the transfer processes, additional means of support have to be provided.

The IANES project focuses on the development of instruments (both methodological and technological) to facilitate the identification and alignment of cooperative work processes in organizations. To promote this research endeavor, three partners (one from the academic sector, two industry partners) join the project and bring together their expertise and already existing instruments. These instruments address distinct aspects of the proposed project aims. Within the project, the instruments are made accessible methodologically and technologically to the other partners in order to identify and implement interfaces and gateways among them to finally provide full support for the IANES processes.

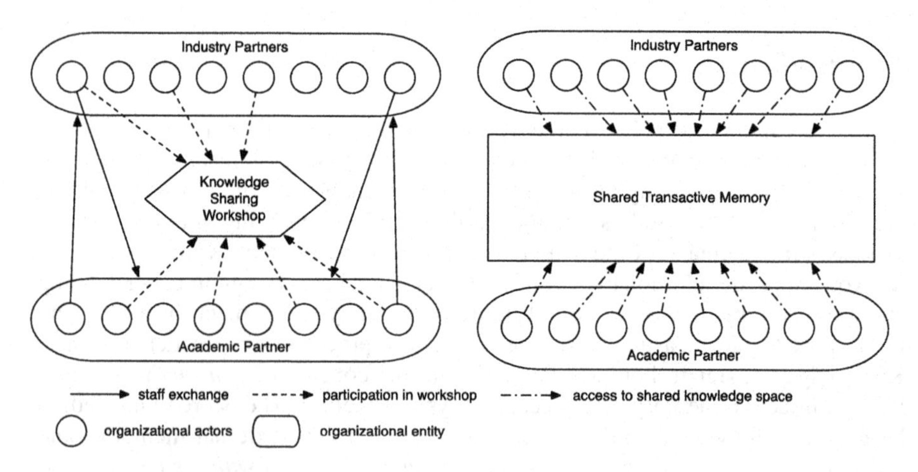

Fig. 1 Co-located, synchronous knowledge sharing (left) and spatio-temporally independent knowledge sharing support (right)

Figure 1 visualizes traditional means of knowledge transfer in industry-academia partnership. When exchanging staff and organizing knowledge sharing workshops (on the left) not all relevant people might be reachable,as knowledge needs might not be known in advance. A shared transactive memory (on the right) provides access to knowledge and the relevant actors for all involved people and thus can resolve these shortcomings through direct access facilities at any time. The aim of this paper is to develop the requirements for such an endeavor by revisiting distributed transactive memory systems and describing an approach supporting different aspects of knowledge sharing within the IANES project as an initial showcase.

The online platform Nymphaea supports spatially and temporally distributed knowledge sharing processes in IANES by providing means to both, retaining relevant knowledge and building a distributed transactive memory that facilitates access to this knowledge for all people involved in the project. Nymphaea has been developed in earlier research of one of the project partners in an eLearning-context [1] and will also be deployed in an extended version in the IANES project itself, making the project a recursive showcase of a part of its own methodology.

In this experience report, we focus on the Nymphaea platform as a means for spatially distributed knowledge sharing. First, the requirements on building a distributed transactive memory system are derived from literature. We then describe the toolset and its support for different aspects of knowledge sharing within the project. Finally, we report on our initial findings on the effects of the approach within IANES.

2 Requirements Supporting Distributed Knowledge Transfer

The challenge of building distributed transactive memory systems to facilitate knowledge sharing beyond a directly interacting group level has initially been addressed by Nevo & Wand [7].

Nevo & Wand claim that transactive memory in distributed settings can be supported by an organizational memory system providing access to:

- Role knowledge – knowledge that is required by definition to take a certain role (e.g. knowledge about how to write program code in a specific language for application developers)
- Instance knowledge – knowledge a person has but which would not be required by his or her formal role (e.g. experiences in supporting international research projects for a secretary)
- Transactive knowledge – knowledge about how to effectively extend one's knowledge by interacting with others. This includes:
    - Conceptual meta-knowledge (ontological concepts needed to describe a knowledge domain)
    - Descriptive meta-knowledge (information about role or instance knowledge, like author, scope, format or creation date)
    - Cognitive meta-knowledge (knowledge about one's own knowledge and abilities)
    - Persuasive meta-knowledge (knowledge about the credibility and expertise of the source)

In providing support to externalize and access these knowledge types, Nymphaea aims at aiding not only content sharing but integrates role knowledge, instance knowledge and transactive knowledge to form a platform for distributed knowledge transfer via semantically enriched content and communications structures. In the following section the features of Nymphaea enabling capturing and delivery of the different types of knowledge are described. The integration of these features to support distributed knowledge transfer is presented subsequently.

3 Knowledge Representation in Nymphaea

The concepts of the Nymphaea platform (cf. https://nymphaea.ce.jku.at) aim to put people seeking for or being able to provide knowledge in control of the transfer process. It also allows situation-specific communication among the parties involved in the transfer process. In the following sections we review the feature-set of Nymphaea clustered along the different knowledge types described above.

3.1 Representing Role Knowledge and Descriptive Meta-knowledge

Role knowledge is represented in Nymphaea using fine-grain content objects. A content object is a conceptual building block within the knowledge to be represented, such as a definition, an example or an explanation. Instead of using a document-centric approach to provide information, content is split in its fundamental didactical elements. These elements can be flexibly arranged and reused to form representations of role knowledge.

Descriptive meta-knowledge is codified in the navigation structures of Nymphaea as well as directly anchored on content objects. The Nymphaea learning environment provides different "workspaces" for users, each one containing the relevant knowledge representations e.g. necessary for a certain task. It provides meta-knowledge about the domain-specific scope of knowledge, its author and creation date. Content within a certain workspace comprises of modules and hierarchically structured content objects. These content objects are enriched with educational meta-knowledge such as ‘definition’, ‘motivation’, ‘background information’, ‘directive’, ‘example’ or ‘self test’. This didactical meta-knowledge is displayed on the right side on top of each content element (cf. Fig. 2) and can be used in the course of individualization when filtering content according to metadata.

Besides structuring content according to educational and domain-specific meta data, Nymphaea provides means to structure content according to level of details (e.g., Overview – cf. “Überblick” in Fig. 2, Content – cf. “Inhalt” in Fig. 2) allowing learners to retrieve content in the desired granularity.

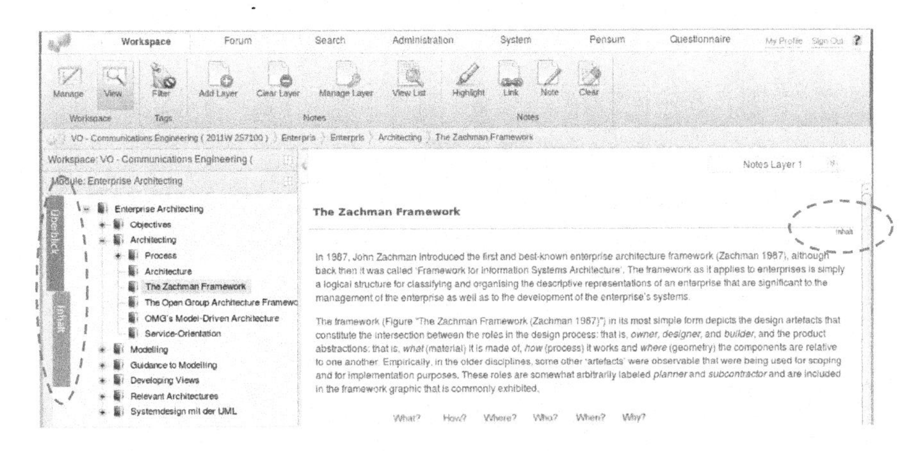

Fig. 2 Content structure

3.2 Representing Conceptual Meta-knowledge

Nymphaea provides an alternative navigation design focusing on domain-inherent structures that can be used complementary to hierarchical navigation. The navigation design represents and organizes conceptual meta-knowledge in a graphical concept map.

Concept maps (cf. [8]) are established means to organize and represent knowledge. They can be used to support the process of eliciting, structuring, and sharing knowledge and aim to enable meaningful learning (cf. [2]). Concept maps use concepts as entity to structure items of interest. Concepts might be central terms, expressions or metaphors, as they represent a unit of information for the person(s) using it. Those items are put into mutual context, leading to a network of concepts. Persons express the items of interest and the relationships by means of language constructs. Per se, there are no restrictions in the naming of concepts or relationships.

Compared to the traditional navigation design, the concept map navigation enables domain-specific and cross-border relationships. Knowledge acquisition paths can considerably differ when using the concept map approach. Instead of implicit learning paths – via hierarchies of modules, learning units, blocks or via internal/external links –, learning paths using a concept map are oriented towards explicit structural relationships beyond hierarchies and domains.

Figure 3 depicts a part of a cross-disciplinary concept map for codified knowledge about 'Enterprise Architecting'. It can be used for navigating learning contents.

Within the map domain-specific associations are used for relating concepts. Furthermore, descriptive meta-knowledge (such as motivation, discussion, etc. - see Fig. 3) is used to semantically describe links between concepts and information

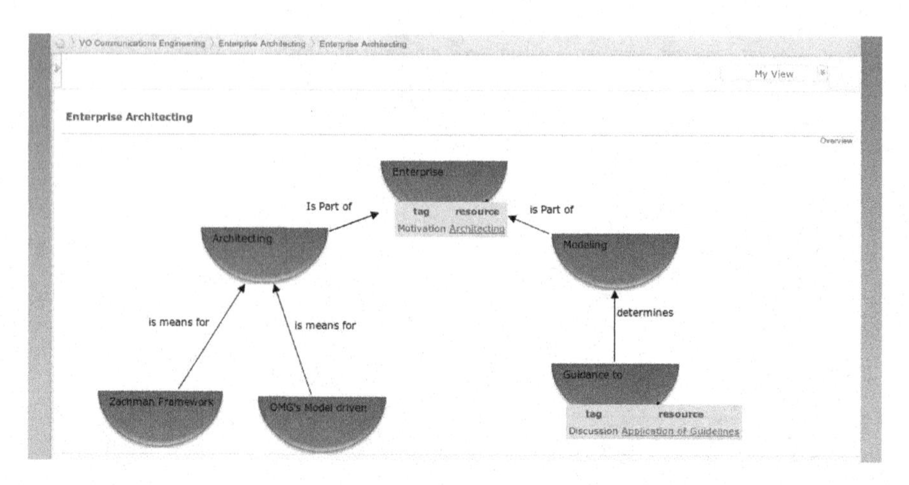

Fig. 3 Examplifying Cmap navigation and content links

resources. Hence, the associative navigation provides additional structural navigation information that shapes learning paths and should guide the individual exploration of content.

3.3 Enabling the Assessment of Cognitive Meta-knowledge

Cognitive meta-knowledge (i.e. knowledge about one's own knowledge) and persuasive meta-knowledge (i.e. knowledge about the credibility and expertise of the knowledge source) are not directly represented in Nymphaea's content structures. Nymphaea rather provides means to enable people seeking for knowledge to assess, whether they need to access and learn certain content. Currently, judging the credibility of the provided sources is not supported.

A specific instrument termed "Intelligibility Catcher" aids the acquisition of meta-knowledge [12]. Intelligibility Catchers (ICs) are grounded in reform-pedagogic and constructivist didactics in general and the Dalton Plan approach in particular [9],[12],[5]]. The main objectives of the Dalton Plan are to provide freedom to learners and enable them to follow individual learning paths, requiring group interaction and collaboration [9][12][4]. The main vehicle guiding learners through larger and complex (group) learning tasks are assignments. For assignments a certain structure has been proposed, which is implemented through ICs. Based on the pedagogic foundation of Parkhurst's assignments, ICs are additionally tailored to be methodically and functionally integrated in online knowledge transfer environments. The IC's structure is as follows (clustered along transactive memory dimensions).

Elements aiding the assessment of cognitive meta-knowledge

- *Preface (Orientation section):* The preface section provides the context motivating the learning tasks.

- *Topic / Objectives:* This section clearly states the central idea of the subject to be learned. This helps learners to stay focused and reflect about their own work on/about this topic.
- *Problems and Tasks:* This section includes all tasks learners work within the frame of the current assignment. It is advisable to state here which problems are to be solved individually by each learner and which problems are to be solved in a group of learners.

Elements aiding the creation of instance and transactive knowledge

- *Written Work:* This section identifies the documentation to be provided by learners. When finished, the involved people discuss written work within a meeting/conference (see below). ICs particularly include references to functionality provided by the platform in order to support tool usage.
- *Memory Work:* In this part of the assignment the intellectual and cognitive work is described. It comprises the intellectual effort to be spent when explorating content and apply it in a reflective way for problem solving.
- *Conferences / Meetings:* While learners are required to manage their own (learning) time, it is often advisable to schedule (online) meetings and check the intermediate progress.

Elements containing descriptive meta-knowledge

- *References:* All additional content for which it is advisable to be read, are referenced here.
- *Equivalents:* The estimated effort (in hours of work) for the assignment is provided here.
- *Bulletin Board:* A forum dedicated to discussions related to the assignment is provided.

Using this structure, guidance on how to use and interact based upon content for knowledge acquisition is provided to potential learners. Following a self-regulated learning paradigm [11], cognitive meta-knowledge is not directly captured and represented in the platform, but users are enabled to self-assess their existing knowledge and learning requirements. Elements specifically targeting at didactically reasonable interaction among learners and with knowledge providers facilitate building transactive knowledge and allow making use of instance knowledge within and external to the platform.

3.4 Capturing and Sharing Instance-Knowledge

Support for individualization of content is provided in Nymphaea with respect to capture instance-knowledge (cf. [1]). Annotations enable individuals to (i) annotate or alter a specific content element, (ii) post questions, answers or comments directly anchored on content and (iii) additionally link the contribution to a discussion theme from the system's global discussion board. The latter link (being part of navigation) guides users to the adjacent discussion of the learning material. In case of real-time online connections, e.g., chats, the questions and answers can pop up immediately on the displays of all connected users (available in a buddy list). In addition, the content elements referred to can be displayed at the same time.

Annotation support for content is realized using a view concept. As soon as provided content is displayed a view is generated like an overlay transparency. The view is kept for further access and reloaded when the content is accessed again. Within a certain view learners can (i) highlight, (ii) link, (iii) add remarks to content elements (compare "How could BPMN...." in Fig 4). The features for view management (add view layer, delete view layer, share view layer, show available views) as well as those for annotations are located in the ribbon-bar at top, whereas the selection of a certain view is provided at the right hand top of the content area (compare "MyView" in Fig. 4).

While annotating content, learners can add internal and external references to content items. Internal references are links between content and communication items, such as entries in the discussion forum or Infoboard, which support context-sensitive discussions. Furthermore, internal links might refer to other elements within the same or a different module. The corresponding features have been included into the annotation icon bar (see Fig. 4 'Link'). Editing internal links requires marking a position in the text that should represent the link. After evoking the respective function located in the ribbon bar at the top a tree with the node of the currently addressed module is displayed. It allows users to select the target of the link (e.g., a forum entry or another content item).

Coupling content and communication is core concept in Nymphaea to support the sharing of instance knowledge. Features supporting sharing are integrated with the individualization features to comprise the possibility to contextualizing individual interactions by directly anchoring them on content elements. Sharing of individual views or creation of shared views, as suggested by Chang et al. [3], is enabled in the system.

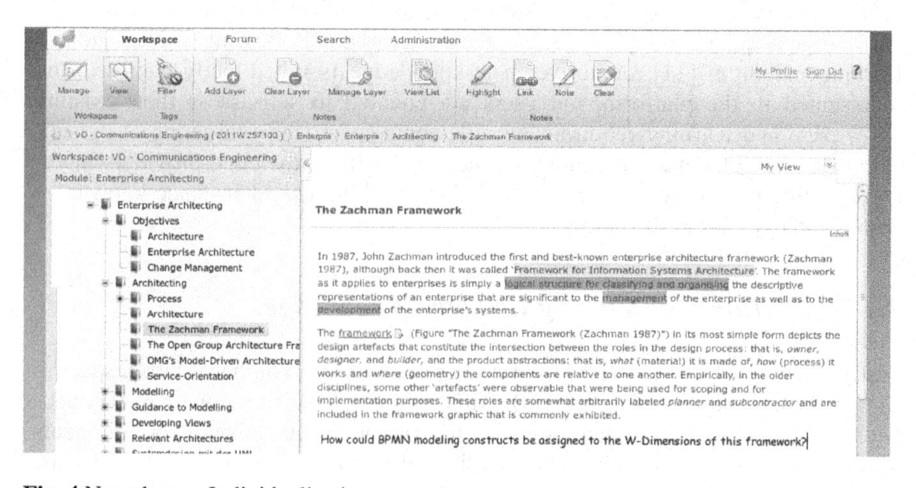

Fig. 4 Nymphaea – Individualization support

The system allows linking content elements to forum and discussion entries entries, and vice versa. Sharing these links in a group enables the group to discuss provided content in context. This feature is particularly useful when users are not only "passive" recipients of content but also actively provide or augment role or instance knowledge in the workspace. Having the discussion documented in the forum provides new users with justifications and background information that has led to previous revisions of content [14].

4 Putting Organizational Memory Support to Practice

The use of Nymphaea to support transactive organizational memory is shown on an IANES-based example in this section. The supporting properties of Nymphaea are also visualized in Figure 5. Figure 5 shows how people can use to system to provide, contribute to and/or access and make use of different types of codified content to ultimately participate in transfer of knowledge. The example is used to map the features of the platform to the support of capturing and making use of different knowledge types that are put into context of a knowledge transfer process. Comments on the covered aspects of transactive knowledge are put in *italics*:

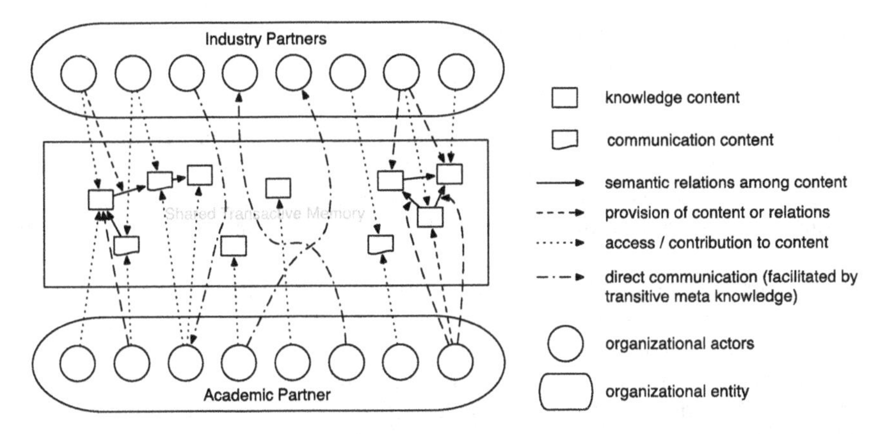

Fig. 5 Nymphaea acting as a shared transactive memoryIn preparation for knowledge transfer, the researcher puts content about concept mapping in general, the chosen variant of the mapping methodology and information about how to use tool support into the Nymphaea platform. *This content is considered role knowledge for people deploying the developed instrument.* She adds didactical content types to visualize the intended use of the content and hierarchically structures the content in a course *(→ descriptive meta knowledge).* Additionally, the systems keeps track of metadata such as author, revision and people having access to the content *(→ descriptive meta knowledge).* In a further step, she provides a concept map about the relationships of the included content, e.g. visualizing which tool features support certain aspects of the methodology to be deployed *(→ conceptual meta knowledge).* Finally, she generates an IC (Intelligibility Catcher) utilizing her content to provide knowledge about what she expects people consuming her content to already know and should finally understand in the course of learning *(→ cognitive meta knowledge),* how their insights should be documented and where additional sources for information within the platform or externally can be found *(→ support for creating instance knowledge).*

A researcher from university has developed a prototype tool for supporting co-operative concept mapping. The industry partners want to deploy this tool in end user studies involving some test customers and thus require knowledge about how the methodology works and in which ways tool support is provided. As development is still ongoing, the research partner frequently publishes tool updates and is highly interested in experience reports and feedback by the industry partners.

In the course of working with the content (via the IC) and deploying the toolset in practice, *instance knowledge* is created through reflecting on the content and practical experiences, and can be integrated with already existing content in the platform by both, the industry partners (learners) and the research partner. Using the annotation feature, comments on shortcomings of the toolset and potential workarounds can be shared with other learners and the researcher. Being anchored directly on the content element e.g. describing an incomprehensible feature, *instance knowledge* is put in and accessible in its context of use. Using discussions or chats, which are equally anchored on content elements, learners can contact each other and the researcher to discuss issues and questions synchronously and asynchronously, thus making direct use of the *transactive knowledge* codified in the platform.

5 Conclusions and Future Work

The Nymphaea platform described in this paper aims at providing organizational memory support in spatially distributed research and development settings with an holistic approach to knowledge capturing and delivery as proposed by Nevo & Wand [7]. Nymphaea is deployed in the FP7-founded project IANES as an organizational memory and is used to internally disseminate knowledge among the partners.

By the time of writing, IANES has been operative only for several months. The platform is already setup and filled with exemplary, yet project-specific content by the academic partner in line with the content creation process described above. Initial feedback of a limited set of users from the industry partners is promising, a formal evaluation, however, is to be conducted later in the project. Empirical evidence on the feasibility and effectiveness of different features for knowledge transfer support in Nymphaea has already been published earlier (cf. [6],[10]). Nymphaea is likely to be integrated in the IANES toolset, as it is intended to act as the central entity for knowledge transfer and interaction support.

Future work will focus on the validation of the suitability of the conceptual framework deployed here to support knowledge transfer in spatially distributed settings. The effectiveness of the toolset will be validated within the project and in external test cases.

A major challenge in this respect is how to measure the impact the Nymphaea platform has on knowledge sharing within the project. IANES will require ongoing use of the sharing platform, so that several usage instances can be examined. Currently, different settings in both, technological and methodological knowledge sharing needs with transfer in both directions (industry – academia) are envisioned (e.g., knowledge about the modeling frameworks of the industry partners that is required at the academic partner for augmentation or knowledge about qualitative empirical research provided by the academic partner and required by the consultants of the industry partner in the course of evaluation). Methodologically, the transactive memory approach will be used as a starting point to derive metrics for measuring the platform's impact. As transactive memories do not focus on plain knowledge consumption but rather on creation, contribution and collaboration, the amount of knowledge shared in the different knowledge categories (to be distinguished by the use of the different Nymphaea features) might provide a quantitative indicator of the platform's impact on knowledge sharing.

The impact of the platform in terms of outcome (i.e. effectiveness and efficiency of knowledge transfer) can be measured by assessing the learning effects of the involved people as well as by reviewing the work processes and outcomes the transferred knowledge was necessary for. Metrics and methodology of measurement are still open here and are subject to the next steps of research in the project.

Concluding, this paper has presented an approach for spatially and temporally distributed knowledge sharing in an industry-academia cooperation based on the concept of transactive organizational memory. The online platform Nymphaea has been described as a means to implement a transactive organizational memory system and facilitate sharing of not only role-specific knowledge, but also person-specific (meta-)knowledge. The application of the approach has been exemplified for IANES, where the system will be deployed, further developed and evaluated over the next years.

Acknowledgments. The research leading to these results has received funding from the European Commission within the Marie Curie Industry and Academia Partnerships & Pathways (IAPP) programme under grant agreement n° 286083.

References

1. Auinger, A., Fürlinger, S., Stary, C.: Interactive Annotations in Web-based Learning Environments. In: Proceedings of ICALT 2004. IEEE (2004)
2. Ausubel, D.P.: Educational Psychology, A Cognitive View. Holt, Rinehart and Winston, Inc., New York (1968)
3. Chang, S., Hassanein, E., Hsieh, S.-Y.: A Multimedia Micro-University. IEEE Multimedia, 60–68 (1998)
4. Eichelberger, H. (ed.): Eine Einführung in die Daltonplan-Pädagogik, p. 223. Studien-Verlag (2002)
5. Eichelberger, H., Laner, C., Kohlberg, W.D., Stary, E., Stary, C.: Reformpädagogik goes E-Learning - neue Wege zur Selbstbestimmung von virtuellem Wissenstransfer und individualisiertem Wissenserwerb Oldenbourg (2008)
6. Neubauer, M., Stary, C., Oppl, S.: Polymorph Navigation Utilizing Domain-Specific Metadata: Experienced Benefits for E-Learners. In: Proceedings of the 29th European Conference on Cognitive Ergonomics (ECCE 2011). ACM Press (2011)
7. Nevo, D., Wand, Y.: Organizational memory information systems: a transactivememory approach. Decision Support Systems 39, 549–562 (2005)

8. Novak, J., Canas, A.J.: The theory underlying concept maps and how to construct them. Technical Report IHMC CmapTools 2006-01, Florida Institute for Human and Machine Cognition (2006)
9. Parkhurst, H.: Education on the Dalton Plan. Nabu Press (1923, 2010)
10. Oppl, S., Stary, C.: Effects of a Tabletop Interface on the Co-construction of Concept Maps. In: Campos, P., Graham, N., Jorge, J., Nunes, N., Palanque, P., Winckler, M. (eds.) INTERACT 2011, Part III. LNCS, vol. 6948, pp. 443–460. Springer, Heidelberg (2011)
11. Oppl, S., Steiner, C., Albert, D.: Supporting Self-regulated Learning with Tabletop Concept Mapping. In: Interdisciplinary Approaches to Technology-Enhanced Learning. Waxmann Verlag (2011)
12. Stary, C.: Intelligibility Catchers for Self-Managed Knowledge Transfer Advanced Learning Technologies. In: ICALT 2007, pp. 517–521 (2007)
13. Wegner, D.: Transactive memory: A contemporary analysis of the group mind. Theories of Group Behavior 185, 208 (1987)
14. Weichhart, G., Stary, C.: Collaborative Learning in Automotive Ecosystems Digital Ecosystems and Technologies. In: 3rd IEEE International Conference on DEST 2009, pp. 235–240 (2009)

Nonlinear Time Series Analyses in Industrial Environments and Limitations for Highly Sparse Data

Emili Balaguer-Ballester

School of Design, Engineering and Computing
Address: Talbot Campus, Poole, Dorset, BH12 5BB
Bournemouth University, UK

Abstract. This work presents case studies of effective knowledge transfer in projects that focused on using nonlinear time series analyses in varied industrial settings. Applications, characterized by intricate dynamical processes, ranged from e-commerce to predicting services request in support centres. A common property of these time series is that they were originated by nonlinear and potentially high-dimensional systems in weakly stationary environments. Therefore, large amount of data was typically required for providing useful forecasts and thus a successful transfer of knowledge. However, in certain scenarios, classifications or predictions have to be inferred from time windows containing only few relevant patterns. To address this challenge, we suggest here the combined use of statistical learning and time series reconstruction algorithms in industrial domains where datasets are severely limited. These ideas could entail a successful transfer of knowledge in projects were more traditional data mining approaches may fail.

1 Introduction

With the increasing possibilities of accessing massive amount of data, prediction of customer's behaviour in industrial domains has become a very active research area during the last twenty years. For instance, one of the most important goals within data mining in Web environments consists of predicting interesting characteristics of users using their previous navigation patterns or *clickstreams* (Martin et al., 2004). These kinds of data analyses enables to better profile the customers commercial Web portals, and in turn, to customize the interface by providing the most suitable services for reach individual user in advance (Fu et al., 1999; Carberry, 2001; Martin et al., 2006). Therefore, it is particularly relevant to anticipate users' demands. Recently, it was shown that powerful nonlinear algorithms for predicting of users' preferences can significantly increase the effectiveness of large web portals of organizations or institutions; provided we have access to a significant amount of historical data (e.g. Martin et al., 2004; Martin et al., 2006, 2007).

R.J. Howlett et al. (Eds.): *Innovation through Knowledge Transfer 2012*, SIST 18, pp. 51–60.
DOI: 10.1007/978-3-642-34219-6_6

Besides web mining, time series analyses have been successfully applied in other industrial settings. For instance, a recent study addressed the more efficient management of Support Centres (Balaguer-Ballester et al., 2008). Support Centres (SCs) usually deal with all the requests reported by either external customers or internal users of a company. The formal contract between the service provider and the customer (the service recipient) usually contains clauses that economically penalize or incentive the SC depending on its performance. Therefore the successful anticipation of service request is an essential aspect in the efficient management of both human and technological resources that are used to solve these eventualities; and thus is of highly significant economical relevance.

Another domain in which predictive analyses have been successful is targeted marketing. The latest marketing trends focus on maintaining and optimizing the behaviour of loyal customers rather than getting new ones i.e. to increase the *net* value of the customer in the long term or lifetime value, LTV (Reichheld, 2001). The relationships between a company and its costumers follow a sequence of action-response cycles, and customer's behaviour typically evolves according to the marketing actions (Pfeifer and Carraway, 2000). For this reason, in targeted marketing, is also essential to predict which subsequent marketing action (for instance, offering a discount to a particular customer) will result in an increase of customer's LTV (Gomez-Perez et al., 2009).

The case studies outlined above are good examples of successful knowledge transfer between Academia and Industry. The algorithms developed in a scientific environment were then implemented in several commercial software Java© platforms and used for improving the performance of the web sites, incidences management centres or marketing policy of the companies, respectively. The processes underlying such applications were typically nonlinear and therefore very challenging for prediction models (Balaguer-Ballester et al., 2008). Nevertheless, forecasts were still possible because large amount of data was available in all situations, enabling us to robustly optimize the parameters of sophisticated machine learning approaches; and thus knowledge transfer was successfully implemented (Martin et al., 2006).

However, in certain challenging scenarios, predictions have to be inferred from very short time series, for instance in biomedical applications using functional magnetic resonance brain-imaging, where the sampling rate is very low and therefore only tens of data patterns are available. Moreover, sometimes the number of accessible variables does not suffice in accounting for the real complexity of the system. In these challenging research situations, the underlying system dynamics is not accessible from the available data. Therefore knowledge transfer may not be successfully implemented using standard machine learning and computational statistics approaches. In this work, we propose the combined use of statistical learning and time series reconstruction algorithms for analysing intrinsically nonlinear problems in industrial settings, particularly when the number of relevant temporal patterns is severely limited. These approaches could be useful in real-life industrial domains were more traditional inferential approaches are not easily applicable due the complexity of the system (many potentially independent variables) and data limitations.

2 Case Studies and Discussion

2.1 Web Mining and Viability of Recommender Systems

The first case study of this work consists of designing recommendation strategies in commercial or institutional web portals using predictive models.

Personalized recommender systems has been a very active field of research during the last decade (Carberry, 2001), and they become an important component of e-commercial Web sites (for instance, Amazon.com). Such systems have been extensively tested and can reliably process large amount of data (Zukerman & Albrecht, 2001). Collaborative filtering is perhaps the most popular approach. This type of systems computes indicators of interest such as frequency of access or indexes of user similarities; and then suggests the most suitable services or products on-line to the user. Some early and well-known examples are Recommender (Hill et al., 1995) or NetPerceptions (Resnick et al., 1994). Those "coarse-grained" analyses have the advantage that they do not need any complex object representation, but only aggregated statistics. Therefore, collaborative filtering is a good strategy in commercial portals provided that "people-to-people correlations" i.e. variations in tastes, accounts for the majority of the variation in user's preferences (Schaffer et al., 1999). In contrast, content-based approaches are based on users' past preferences in order to forecast his/her future behaviour; thus, those learning algorithms are particularly suitable for situations in which users exhibit a predictable behaviour i.e. a significant "item-to-item correlation". Figure 1 shows a simple schema of these two approaches.

An example of the successful applications of recommender systems to commercial portals is the Java platform iSUM©; developed by the Spanish IT and software company TISSAT© (www.tissat.es). The data mining capabilities of this product were entirely based on the results shown in Martin et al. (2004, 2006, 2007). The two classes of recommender algorithms outlined above were implemented, in order to provide the optimal solution depending on data characteristics. For instance, in Martin et al. (2004), content-based recommenders based on Associative Memories, Time-Delayed Perceptrons (Balaguer-Ballester et al., 2002), Classification and Regression Trees (CART; Duda et al., 2001), and Support Vector Machines (SVM; Vapnik, 1998) were used for predicting the services related to bureaucracy, shopping or entertainment etc. based on the historical information of individual users. The optimal models where able to predict with an 80% of success the preferred service by users in a large institutional portal; thus validating the "item-to-item" assumption in this scenario (Martin et al., 2004).

In other situations however, "people-to-people" collaborative recommendation is more appropriate, since the aim is to find inter-user similarities rather than idiosyncratic behaviours of individual users. In particular, in Martin et al. (2006, 2007) users' behaviour was profiled by using different neurologically inspired clustering algorithms such as Adaptative Resonance Theory-2 Networks (ART2, Carpenter & Grossberg, 1991); which have been designed to adapt clusters to new data patterns, without disrupting the already established clusters. This feature can support on-line tracking of user profiles. Moreover, in this work, we developed an artificial data generator; which permitted to benchmark different recommendation algorithms in arbitrarily complex web scenarios. Our approach was able to accurately find groups of similar users on a range of real and simulated e-commerce websites; and afterwards new users were offered services which were primarily accessed by the users of the same group. As a result, this approach improved by 1.5-2.5 times the prediction accuracy obtained with more standard approaches (Martin et al, 2007). As briefly mentioned earlier, this functionality was transferred to the web platform iSUM©. The recommender systems above described were integrated in the Artificial Intelligence Module, available in versions 4.x and superior of such software product. These novel functionalities permitted to successfully position iSUM© within the international market of intelligent web mining software.

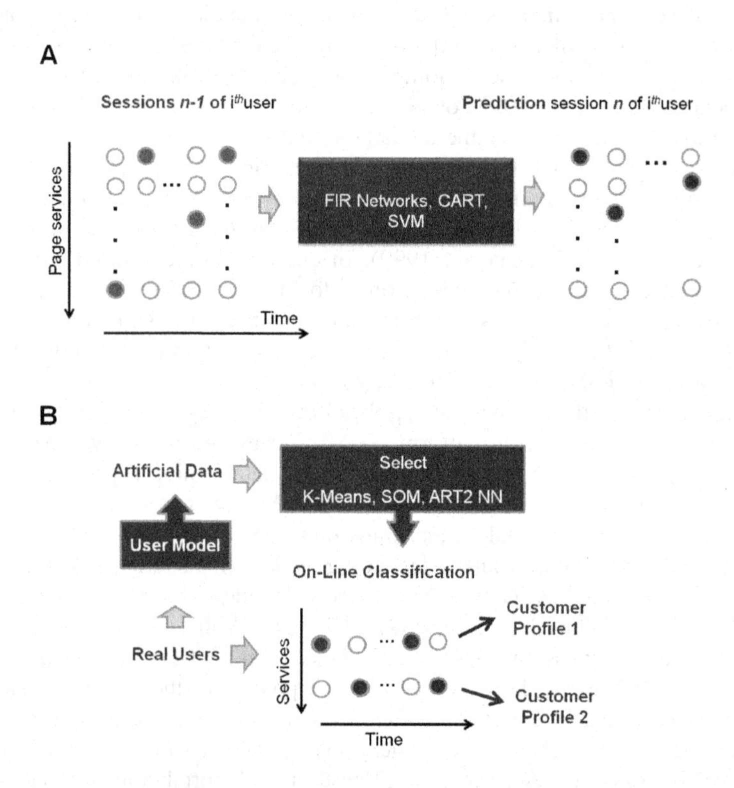

Fig. 1 The most widely used recommender systems. **A**. Content-based recommendation, where "item-to-item" correlation is significant and thus prediction of the future preferences of customers is feasible (FIR NN: Finite Impulse response Networks; see e.g. Balaguer et al., 2002). **B.** Collaborative filtering, where the aim is to find inter-user similarities rather than idiosyncratic behaviours of individual user, thus unsupervised clustering algorithms are typically used (SOM: Self-Organizing Map, Kohonen, 1997). In both approaches (especially in Content-based-approaches), accuracy for short time series is severely compromised.

Unfortunately, these two recommendation algorithms require collecting large amounts of data from each user in order to reliably estimate the parameters of such nonlinear models. This requirement is of course not exclusive of e-commerce websites but is common to many predictive scenarios in other industrial domains; as discussed in the following sections.

2.2 *Predicting Service Requests in Support Centres*

A second example of successful predictive modelling in industrial domains is the automatization of Support Centres management.

Although telephone has been the traditional way to provide support, nowadays the Internet is widely used (Balaguer-Ballester et al., 2008). The contracts between service provider and customers (service level agreement) define a series of quality measures or service level measurements (SLMs) that are used to evaluate the quality of the support service. Different system performance and business parameters can be considered for SLMs: aggregated statistics of time-dependent parameters, service availability, number of affected users, metrics based on a particular business process, etc. These measurements are automatically collected, maintained and analysed in order to manage the service support process. The "Help Desk" platform consists of several software applications that allow recording all the information involved in a SC; which is considerably wide: general information, purchases, complaints, etc. SCs are especially interested in managing these forthcoming events as good as possible. Moreover, is preferable that, whenever it is possible, problems are solved by operators at the first levels of the system hierarchy. Therefore the prediction of the number of forthcoming service requests as well as the time when they occur will permit to optimize the resources used to solve these eventualities.

For instance, in Balaguer-Ballester et al. (2008) we analysed different eventualities managed by an official SC (CETESI©; a project which aimed for a complete integration of institutional services provided by local authorities in Valencia, Spain). In this project, we used time series reconstruction algorithms in order to evaluate the complexity of the system dynamics, which varied during the four years of recordings. Therefore, we were able to determine when a linear predictor will suffice (for low-complex multivariate time series of incidences) or rather highly-parameterized nonlinear approaches where more suitable. Then, we selectively applied linear/nonlinear models (auto-regressive moving averages with exogenous inputs and neural networks) depending on the previous analyses in order to apply the most robust and reliable method to predict future events in each situation.

Using this "meta-approach" for selecting the most suitable model, we obtained an 86–94% of out-of sample prediction accuracy in certain management projects; contributing to the successful fulfilment of the service level agreement. As a result, local authorities renewed the contract with TISSAT©; yielding to an increase in the company's revenue and to a better positioning in the Support Centres Management sector.

Nevertheless, those nonlinear time series predictions were less accurate in the projects were long time series (two years of recordings) were not accessible (Balaguer-Ballester et al. 2008).

2.3 Optimizing Strategies in Marketing Campaigns

The most common strategy for increasing the loyalty of customers is by offering them the opportunity to obtain some gifts as the result of their commercial transactions. For instance, the company can assign "virtual credits" for each purchase of those products which wants to be promoted. After a certain number of purchases, the customers can exchange their virtual credits for the gifts offered by the company.

Therefore, the policy of this targeted marketing campaign consists of establishing the appropriate number of virtual credits for each promoted item. The goal is to achieve an optimal trade-off between the cost of the marketing campaign (for instance, the value of the gifts) and the increase in the amount of purchases as a result of the campaign. Commercial domains of this kind can be viewed as a Markov decision problem, in which a company decides what action to take given the current customer state (Abe et al., 2004).

This is however a difficult problem, due to the large number of variables that can be potentially used for characterizing the state of the customer. In Gomez-Perez et al. (2009), we profiled the behaviour of the customer using three optimal features in Relational Marketing which are the last transaction date a.k.a. Recency, Frequency and Monetary value (Figure 2). After that, we used reinforcement learning algorithms (RL, Sutton and Barto, 1998) in order to determine the optimal marketing policy i.e. the assignment of the appropriate number of virtual credits to each customer for each purchase.

Moreover, targeted marketing can have a very complex characterization of the transactions that are involved. This high-dimensionality requires the combined use of RL algorithms with some dimensionality reduction approaches. For overcoming this drawback, we combined Q-Learning methods with SOM networks (Smith, 2002), which enabled us to compute the optimal policy in a complex credit assignment problem (Gomez-Perez et al., 2009; see Figure 2).This combined approach indicated that incurring in more marketing costs for certain customers (some of the most loyal ones) results in a significant increase (over 50% in some cases) of the long-term LTV with respect to the previous strategy (established manually by marketing experts).

Therefore our algorithm revealed that the company's policy had a lack of exploration of apparently non-profitable marketing actions in the short term, which results in larger long-term benefits. Nevertheless, for other groups of customers the manual strategy could not be largely improved suggesting that, in spite we have about 10^6 data patters, a much large amount of data was still required for optimizing the marketing policy on such customers.

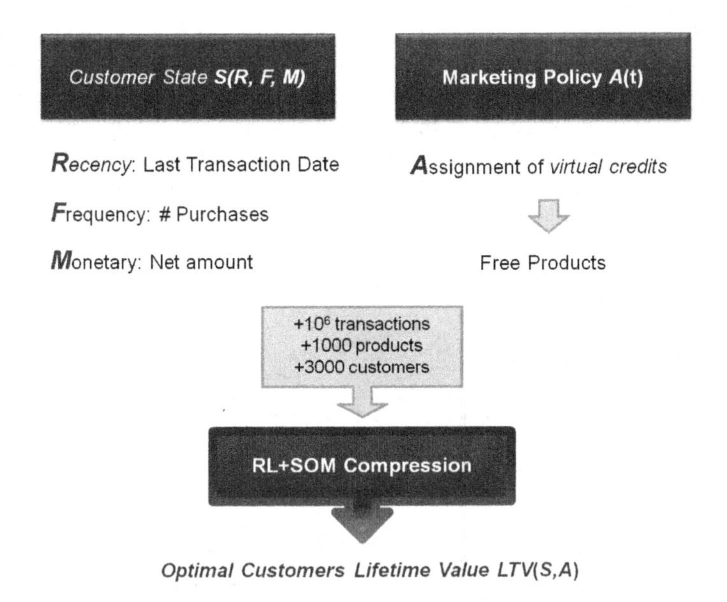

Fig. 2 Methodology for finding the optimum policy $\boldsymbol{A}(t)$ of a Marketing campaign, that is the sequence of marketing actions which maximizes the predicted customers long-term life time value (LTV). A reinforcement learning algorithm (RL, in this case Q-learning) combined with a SOM for compressing the variables space (consisting of customer states S and marketing actions) enabled us to improve the company benefits (Gomez-Perez et al., 2009).

2.4 Limitations of Standard Approaches in Short Time Series

The element that is common for all three presented above case studies is the availability of large number of "independent" variables and, simultaneously, of large amount of data. Therefore, the underlying dynamical process, which was generating the observable temporal patterns, could be inferred using machine learning approaches. Nevertheless, these algorithms were in general less successful in improving the expert-supervised strategies when long time series where not available.

The reason of this failure can be understood in the light of nonlinear time series analyses theorems (Sauer et al., 1992): When the number of variables in the system is not sufficiently large, we cannot precisely determine the underlying dynamics. For instance, trajectories corresponding to two different customer's behaviours may look tightly entangled in a two dimensional space of recorded variables (x-y plane in Figure 3), and therefore two truly separated trajectories exhibit multiple crossings. In those ambiguous patterns (crosses in Figure 3), the future direction of the flow cannot be precisely predicted. It is well-known that this ambiguity can be resolved if we add sufficient number of delayed version of our original axes (Takens, 1980); provided we have noise-free and unlimited data. Nevertheless, those assumptions can be particularly inaccurate for short time series.

A possible alternative consists of "supplanting the missing variables" of the system by nonlinear functions of the recorded variables and suitable temporal delays (for instance see Figure 3, z-axis). This simple approach aims to create a sufficiently expanded space such that trajectories are not ambiguous anymore. Therefore, trajectory analyses are feasible by using linear classifiers which have few parameters to estimate and thus may not require large datasets for obtaining robust and generalizable predictions.

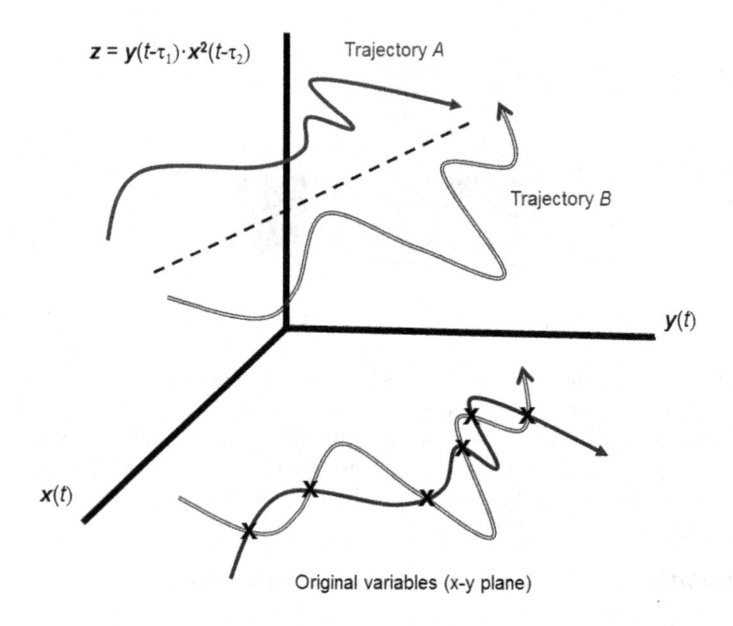

Fig. 3 Reconstructing nonlinear dynamics in noisy time series. When the number of recorded variables is not sufficient (for instance, the *x-y* plane in the figure), trajectories are entangled and thus is particularly difficult to discern whether those trajectories correspond to different classes. The expended space including a nonlinear combination of variables and their delayed version as new axes (for instance the z variable) can disambiguate the system's dynamics, enabling robust classification using simple (low-parametric) models.

For instance, in Balaguer-Ballester et al. (2011), we used simple discriminants (with the aid of *kernel*-representations) for successfully predicting the rodent's behaviour in maze based on neurophysiological recordings. These ideas can be applied to other complex scenarios and thus potentially provide solutions to particularity problematic problems in different industrial settings.

3 Conclusions

This study presents three examples of successful knowledge transfer in very varied industrial domains. Towards that goal, a wide range of machine learning algorithms specialized in temporal patterns processing was used. These approaches helped to improve the existing strategies in e-commerce, supported centres management or targeted marketing scenarios, and others. These novel functionalities in data mining and artificial intelligence were implemented in existing or new commercial software products of TISSAT©. As a result, the company strengthened its position in these strategic sectors. Nevertheless, predictions were much more effective in situations where many system variables and large amounts of data were accessible. This work suggests an alternative approach designed for classification problems based in short nonlinear time series in noisy conditions; which has been recently used in neuroscientific applications. These ideas could help in the successful completion of projects in industrial environments where standard approaches are unfeasible.

References

Abe, N., Verma, N., Schroko, R., Apte, C.: Cross channel optimized marketing by reinforcement learning. In: Proceedings of the KDD, pp. 767–772 (2004)

Balaguer-Ballester, E., Camps-Valls, G., Carrasco-Rodriguez, J.L., Soria, E., del Valle-Tascon, S.: Effective one-day ahead prediction of hourly surface ozone concentrations in eastern Spain using linear models and neural networks. Ecological Modeling 156, 27–41 (2002)

Balaguer-Ballester, E., Lapish, C., Seamans, J., Durstewitz, D.: Attracting dynamics of frontal cortex ensembles during memory-guided decision making. PLoS Computational Biology 7(5), e1002057 (2011), doi:10.1371/journal.pcbi.1002057

Balaguer-Ballester, E., Soria, E., Palomares, A., Martín-Guerrero, J.D.: Predicting service request in support centres based on nonlinear dynamics, ARMA modelling and neural Networks. Expert Systems with Applications 34, 665–672 (2008)

Carberry, S.: Techniques for Plan Recognition. User Modeling and User Adapted Interaction 11, 31–48 (2001)

Carpenter, G.A., Grossberg, S.: ART2: Self-Organization of Stable Category Recognition Codes for Analog Input Patterns. In: Pattern Recognition by Self-Organizing Neural Networks. MIT Press (1991)

Duda, R.O., Hart, P.E., Stork, D.G.: Pattern classification. John Wiley and Sons (2001)

Fu, Y., Shandu, K., Shih, M.: Fast clustering of web users based on navigation pattern. In: Proceedings of SCI 1999/ISAS1999, Orlando, USA (1999)

Gómez-Pérez, G., Martín-Guerrero, J.D., Soria-Olivas, E., Balaguer-Ballester, E., Palomares, A., Casariego, N.: Assigning discounts in a marketing campaign by using reinforcement learning. Expert Systems with Applications 36, 8022–8831 (2009)

Hill, W., Stead, L., Rosenstein, M., Furnas, G.: Recommending and Evaluating choices in a virtual community of use. In: CHI 1995: Conference Proceedings on Human Factors in Computing Systems, Denver, USA, pp. 194–201 (1995)

Kohonen, T.: Self-Organizing Maps, 2nd edn. Springer, Berlin (1997)

Martín-Guerrero, J.D., Balaguer-Ballester, E., Camps-Valls, G., Palomares, A., Serrano-López, A.J., Gómez-Sanchís, J., Soria, E.: Machine Learning Methods for One-Session Ahead Prediction of Accesses to Page Categories. In: De Bra, P.M.E., Nejdl, W. (eds.) AH 2004. LNCS, vol. 3137, pp. 420–424. Springer, Heidelberg (2004)

Martín-Guerrero, J.D., Lisboa, P.J.G., Palomares-Chust, A., Soria, E., Balaguer-Ballester, E.: An approach based on Adaptive Resonance Theory for analyzing the viability of recommender systems in a citizen web portal. Expert Systems with Applications 33, 743–753 (2007)

Martín-Guerrero, J.D., Soria, E., Gómez-Sanchis, J., Soriano-Asensi, A., Palomares, A., Balaguer-Ballester, E.: Studying the feasibility of a recommender in a citizen Web Portal based on user modeling and clustering algorithm. Expert Systems with Applications 30, 299–312 (2006)

Pfeifer, P.E., Carraway, R.L.: Modeling customer relationships as markov chains. Journal of Interactive Marketing 14, 43–55 (2000)

Reichheld, F.F.: The loyalty effect: The hidden force behind growth, profits, and lasting value. Harvard Business School Press, Boston (2001)

Resnick, P., Iacovou, N., Suchak, M., Bergstrom, P., Riedl, J.: An Open Architecture for Collaborative Filtering of Netnews. In: Proceedings of the Conference on Computer Supported Cooperative Work, pp. 175–186. Chapel Hill (1994)

Sauer, T., Yorke, J., Casdagli, M.: Embedology. J. Stat. Phys. 65, 579–616 (1992)

Schafer, J.B., Konstan, J., Riedl, J.: Recommender Systems in E-Commerce. In: Proceedings of the First ACM Conference on Electronic Commerce EC 1999, Denver, USA, pp. 158–166 (1999)

Smith, A.J.: Applications of the self-organising map to reinforcement learning. Neural Networks 15, 1107–1124 (2002)

Sutton, R.S., Barto, A.G.: Reinforcement learning: An introduction. MIT Press, Cambridge (1998)

Takens, F.: Detecting strange attractors in turbulence. Springer lecture notes in mathematics, vol. 898, pp. 366–381 (1981)

Vapnik, V.N.: The nature of statistical learning. Springer, New York (1999)

Zukerman, I., Albrecht, D.W.: Predictive Statistical Models for User Modeling. User Modeling and User Adapted Interaction 11, 5–18 (2001)

Next-Practise in University Research Based Open Innovation - *From Push to Pull: Case Studies from Denmark*

Jens Rønnow Lønholdt*, Mille Wilken Bengtsson, Lone Tolstrup Karlby, Dorthe Skovgaard Lund, Carsten Møller, Jacob Nielsen, Annette Winkel Schwarz, and Kristoffer Amlani Ulbak

Technical University of Denmark, Denmark
lonholdt@gmail.com

Abstract. How do we ensure knowledge transfer from universities in the most effective and efficient way? What is the right balance between a *push* and a *pull* approach? These issues have been discussed at length and various methods of intermediary facilitating and ways to organise the transfer have been tried in different contextual settings at universities all over the world. Lessons learned are mixed and naturally varies from country to country. This paper presents a recently completed development project concerning the transfer facility at the Technical University of Denmark (DTU). The project focused on the pull function and the capacity development of the SMEs as this was the main lessons learned during the initial phase of the project. The paper also presents four Danish innovation projects that illustrate the use of the pull–based concept. Last but not least, the paper presents a new post–graduate education at DTU in design and management of projects in network. It supports competence development within efficient knowledge transfer. Finally conclusions and recommendations will be presented and discussed based on the above six cases within university research based knowledge transfer.

Keywords: Universities, Knowledge Transfer, Open Innovation, Push and Pull, SMEs.

1 Background and Introduction

How do we ensure knowledge transfer from universities in the most effective and efficient way? What is the right balance between a *push* and a *pull* approach? These issues have been discussed at length and various methods of intermediary facilitating and ways to organise the transfer have been tried in different contextual settings at universities all over the world. Lessons learned are mixed

* Corresponding author.

R.J. Howlett et al. (Eds.): *Innovation through Knowledge Transfer 2012*, SIST 18, pp. 61–77.
DOI: 10.1007/978-3-642-34219-6_7

and naturally varies from country to country. However, in the current competitive climate it is crucial to ensure this transfer especially in Europe, facing a massive competition from Brazil, Russia, India and China – the so called BRIC countries. We do see one predominant lesson learned: the demand driven pull approach seems to be the most efficient especially concerning Small and Medium Enterprises (SMEs), which have a large growth potential and which are in dire need for fresh research based knowledge. Consequently, a lot of emphasis should be put on their knowledge transfer.

The established intermediary organisations and functions are causing a growing concern because they tend to be more bureaucratic and not orientated towards a technical and business development profile. Some, including a large number of SMEs, see them as a barrier rather than a facilitator for the open pull-based approach. In addition to this very few SMEs have neither the necessary diagnostic tools for specifying the needed pull nor the capacity for receiving knowledge. To sum up, there is a need for reinventing or reshaping the transfer functions for achieving the right balance between a focal pull function and a supporting push function.

This paper presents a recently completed development project concerning the transfer facility at the Technical University of Denmark. The project focused on the pull function and the capacity development of the SMEs as this was the main lessons learned during the initial phase of the project. The paper also presents four innovation projects that illustrate the use of the pull–based concept:

- A strategic project aiming at developing an urban Danish municipality into a *City of Knowledge.*
- A regional climate change mitigation project driven by the municipalities in the Capital Region of Denmark.
- A regional project using multidisciplinary and untraditional innovation processes to solve in-door climate problems caused by contaminated soil or ground water.
- A municipal Public-Private-Partnership project within the area of steel and metal manufacturing.

Last but not least, the paper presents a new post–graduate education at the Technical University of Denmark in design and management of projects in network. It supports competence development within efficient knowledge transfer. Finally conclusions and recommendations will be presented and discussed based on the above six Danish cases within university research based knowledge transfer.

2 Technical University of Denmark: *ViTiS Development Project 2008-2011*

In 2008, the Technical University of Denmark (DTU) launched a development project named *ViTiS* (In English: Knowledge to Society). The project in addition to its own development activities functioned as an umbrella for a number of external funded development projects related to knowledge transfer and processed their results. They included a variety of relevant internal as well as external co-operation partners. Special focus was put on SMEs based on the previous given logic. As part of the ViTiS Project specific and SME targeted processes, methods and tools for effective and efficient knowledge transfer were developed and tested in real life business situations. Some of the main findings are (DTU ViTiS – Knowledge to Society – debriefing 2008-2012, 2011):

- Universities must see themselves as an integral and important part of a *public–private system* that initiates and stimulates innovation based on research results and knowledge generated at the universities.
- At the same time, they must act as demand driven knowledge *providers* – not only as knowledge *pushers*. Pure administrative oriented match-making units are therefore not fully equipped to support successful knowledge transfer to private companies especially SMEs.
- Internal structures and priorities at the universities have to be reconsidered before starting organised knowledge sharing. Scientists have to be trained to participate in the dialogue.
- Furthermore, university top management has to create incentive structures for the researchers whose main occupation naturally is research and education.
- To support their own development in relation to efficient, effective and sustainable knowledge transfer, universities must participate actively in extensive external networking with outside stakeholders and co-operation partners like business organisations, other universities and public and private institutions. This also provides them with a wider receiving networking system for their knowledge transfer.

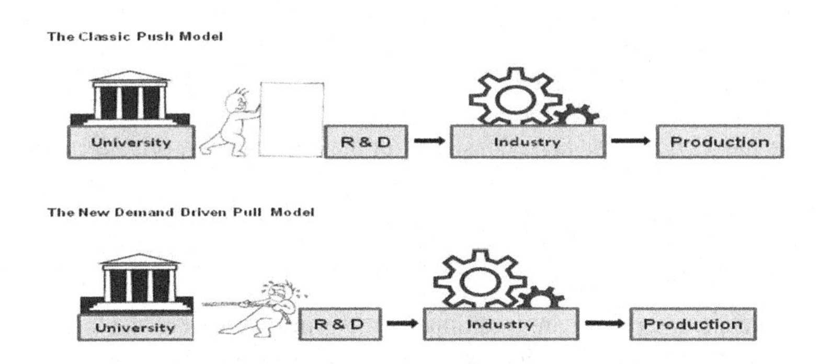

Where to look for barriers?

The project was based on the idea that innovative companies looking for growth and development will benefit from *pulling* needs based knowledge from DTU. The exchange is expected to create viable and beneficial collaborations for the companies as well as the university. Barriers are sometimes generated by factors and systems outside the control of the university. This includes the Danish comprehensive innovation and business development support system, which comprises public as well as private entities and which from the point of view of the SMEs is mostly regarded as complicated, not too transparent and time consuming. However, as the most severe barriers are often found within the university itself, ViTiS placed a considerable focus on the internal system and its competences.

As for organisation of the project a small group of permanent staff proved to be a cost effective and flexible way to create close connections within the university, with other academic and public and private partners outside the university, and especially with the SMEs. The highly skilled and experienced staff was well versed in research as well as business development. Moreover, the project had top management support as it was referenced to the Vice President of the University, which strongly legitimised the project both internally and externally and helped facilitate implementation and co-operation.

Link: www.annefjs.dk/webOSSS/

Link: http://www.dtu.dk/Erhvervssamarbejde.aspx

Case No 1: Municipality of Lyngby-Taarbæk: *City of Knowledge and Urban Creativity 2010 – 2020*

In 2010, the Municipality of Lyngby-Taarbæk launched the idea of becoming a *City of Knowledge and Urban Creativity*. The idea was based on the fact that the municipality hosts a substantial number of creative and knowledge based companies including DTU. It is situated in the vicinity of the capital Copenhagen with a population of approx. 53,000 inhabitants. In collaboration with the City Council, the Committee of Commerce prepared an ambitious strategic plan. The aim was to develop Lyngby-Taarbæk into a green and sustainable university town developing and promoting welfare innovation and municipal services to businesses and citizens.

A strategy with nine focal points

The basic model is a *triple helix* consisting of the municipality, private companies, DTU and other educational institutions. The model was created to generate innovative synergy between *providers* of knowledge, *consumers* of knowledge, regulatory *authorities* and private companies. This public private innovation model faces several challenges:

- Citizens i.e. tax payers are traditionally the primary focus of the local Danish politicians and administration. Not companies. Consequently it was quite new for a Danish municipality to strategically address private companies and their commercial potential.
- Comprehensive public strategic projects with a large number of partners risk too slow a progress seen from the perspective of private companies and therefore the private companies tend to lose interest fairly fast.

However, momentum and interest was kept through the process as focus was fully on the above two issues. The strategy with its *nine focal points* (see below) was politically approved by the Municipality at the end of 2011.

The nine focal points of the City of Knowledge Strategy

The three dark green boxes indicate the main action lines of the strategy. The aim of the strategy was to develop a coherent and a sustainable urban municipality supported by setting up new services and qualities. The strategy comprises a variety of action points. The green boxes might separately be focal areas in local business development plans.

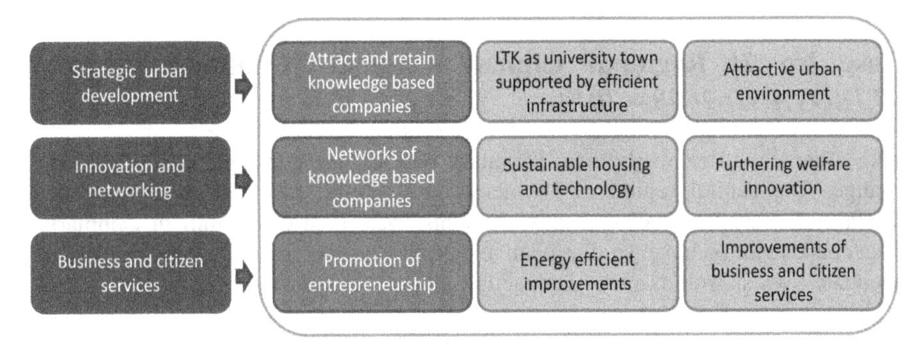

The role of *DTU MatchPoint*

To boost the strategic planning process in terms of technology transfer, DTU fairly early in the strategic process set up a satellite unit called *MatchPoint* in the centre of the city. This was done to ensure practical promotion of the collaboration between the university and private companies especially SMEs. The main public library was used as a venue for this new unit and it became a meeting place for the SMEs to come and present their knowledge need for business development. Here they were met by not only the municipal business development services but also the university matchmaking unit.

The unit began its task with an analysis of business sectors, visits to selected SMEs and open events on topics of interest to the stakeholders. A modern digitized public library certainly has a role to play as an information provider to industry via the local business development service. But based on the results and experience of *MatchPoint*, the role of the university is not so evident. Actually, DTU has drawn the conclusion that the university cannot act as an operator, but rather as sub-supplier of science based knowledge from the research departments. Consequently the match making unit has been internalised in the municipal administration.

Challenges ahead

The municipality is now facing the critical issue of how to go forward with the implementation and keep local companies on board as active players rather than objects for the process. The municipality is highly dependent on contribution of the companies to the development activities. The companies – especially the

SMEs – might have a weak awareness of the initiative and thus give it low priority because short-term advantages are not evident. So how do you create a *pull interface* in relation to implementing the strategy?

A possible solution, which is investigated presently, is to complement the strategic long-term activities with a range of small practical projects. Their objective should be to demonstrate viable and beneficial results within areas of interest for the companies. Furthermore, initiatives to simplify municipal administration in relation to private companies or improve local business services are under way and are expected to encourage companies to participate actively.

Link: http://www.e-pages.dk/vidensbystrategi/4/

Case No 2: Regional Climate Change Mitigation Project: *KLIKOVAND 2008 – 2014*

Like other countries worldwide, Denmark has felt the consequences of climate change. The capital region of Denmark has suffered recurrent heavy summer rainfalls that are unequalled to any other historical data on rainfall in Denmark. The consequences have so far been extremely costly flooding due to lack of adequate flow as well as storage capacity in the sewer systems.

Unique municipal cooperation across water boundaries

When it comes to handling flooding and climate change, the municipalities are the responsible authorities in Denmark. Around 50,000 hectare of land is considered at risk of flooding because of increased water levels and heavy rainfall. The Danish Environmental Protection Agency and the Danish consultancy company Rambøll have estimated that it is necessary to make a yearly investment of around 65 million € until 2040 to expand the capacity of the sewer systems in order to adapt to climate changes and minimise the risk of flooding (Rambøll, 2009).

Based on this and due to a heavy flooding in July 2011, the municipalities and public water companies in the Capital Region of Denmark have formed a partnership, *KLIKOVAND* (Danish abbr. for Climate, Municipalities and Water). The KLIKOVAND partnership has acknowledged the inadequacy of public financing of a renewed sewer system. The premises are that public authorities and agencies, business and civil society all have to develop smart institutional and practical actions in order to respond to climate changes in relation to water (see also EU Commission, 2009 and OECD, 2008).

Therefore KLIKOVAND aims to improve the regional and municipal decision and policy making process in relation to the planning and development of new and existing urban areas in the capital region. The uniqueness of KLIKOVAND also stems from the fact that it is the only Danish project within this area which is completely pull based in relation to the municipalities. This makes it rather innovative in its approach. The figure below illustrates the KLIKOVAND concept including its four main development tracks.

The main aim of the project is to qualify the individual municipal climate change mitigations plans through a structured and co-ordinated process between the participating municipalities taking the specific needs of each participating municipality into consideration and by processing research based knowledge transfer.

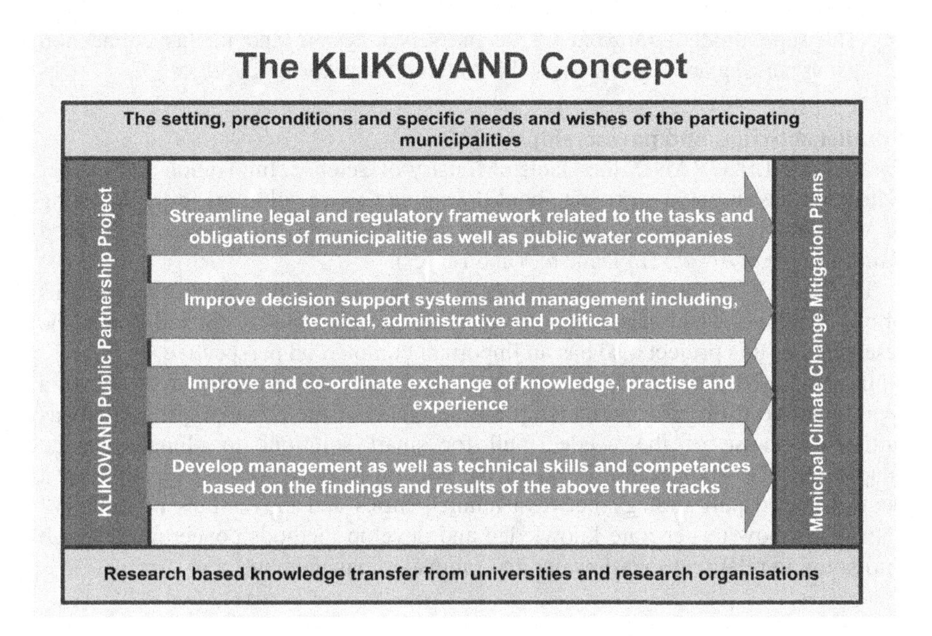

Lessons learned in infancy

So far, the lessons learned are that it requires comprehensive and multifaceted project management capacity with well-developed skills for multi-interest projects to plan and implement this kind of pull based project. It must be remembered that while the project as an idea is quite adult it is still in its infancy concerning implementation. Furthermore, the time needed to promote the project idea and concept and to mature the project environment should never be underestimated. It is also crucial to nurture the motivation of the partners constantly during project implementation since it is a very diverse multi stakeholder project environment with inbuilt conflicts of interests.

Challenges demands new approaches

The challenges facing the municipalities demand new approaches to urban spatial planning and the handling of wastewater and rainwater because it is necessary to change behaviour and get citizens and private companies to be actively involved in meeting the challenges:

- In 2009, the Danish legislation stated that the municipalities had to outsource the management of wastewater from companies and households to independent but public owned entities (Vandsektorloven, 2009). The law only made it possible for the waste water services to charge companies and households for the service of managing waste water and not the necessary investment in adapting to climate change.

- Necessity of *greening* all available surfaces (roofs, walls, streets) in urban areas in order that rainwater will seep into the ground instead of contributing to floods in streets and on public and private properties.
- The separation of rainwater from the sewer system and in this connection using parking areas and sport fields as temporary storing facilities.

Parallel activities and partnerships

Parallel to KLIKOVAND, the Danish Ministry of Science, Innovation and Higher Education has initiated a project about developing smart solutions to the handling of wastewater and increased water level in urban areas (the project is named *Water in Urban Areas*) (In Danish: Vand i Byer).

The partnership includes universities and research centres within the domains of city planning, water supply system and construction industry. In addition to the research part this project also has an important commercial perspective.

In 2011, KLIKOVAND and Water in Urban Areas decided to have a representative from each partnership on the board of the other project. This was initiated because of the needed pull for smart solutions to climate change adaptation. The cooperation between the two projects is the first example of a demand based pull strategy between municipalities and universities in Denmark concerning how to generate knowledge and develop methods concerning climate mitigation in relation to wastewater and rainwater management.

Link: www.klikovand.dk

Case No 3: Capital Region of Denmark In-Door Climate Project: *NYMIND 2011 – 2012*

The NYMIND Innovation and Development Project address mitigation of in-door climate problems in houses or facilities situated on contaminated sites (Amternes Videnscenter for Jordforurening, 2002). The project name is a Danish abbreviation for *New Methods for In-door Climate*. It has been launched by the Capital Region of Denmark because they recognise that the traditional development tracks and venues did not provide sufficient new ideas with adequate innovation power. It was not enough to include the relevant scientific and technical departments from universities, research and development institutions, as well as consulting engineers and construction companies. Actually, the Capital Region of Denmark had prior to the NYMIND Project conducted a similar and traditional innovation project with the said co-operation partners and this did not give satisfactory results.

Consequently, a new approach was required in order to deepen and broaden the innovation perspective. This new approach has two distinct but interlinked dimensions based on the experience with the first project:

1. To seek to widen the innovation perspective through a multidisciplinary approach mainly involving disciplines which are not traditionally involved in solutions related to in-door climate problems
2. To apply untraditional innovation approaches including art based creativity.

An inner and outer core

The NYMIND Project started in August 2011 and is preliminary planned to run until the end of May 2012. The concept is illustrated in the figure overleaf. It shows that the main aim of this project is to create an international innovation platform and identify viable innovation tracks. The platform should identify, formulate and organise new innovation tracks leading to new solutions for in-door climate problems. The inner core is a group of multidisciplinary scientists with their field of expertise outside the area in question, supported by an outer core of scientists much like themselves. In order to ensure that the innovation tracks formulated by this inner core is realistic, viable and implementable, these *non-topic scientists* are *guarded* by topic related scientists that screen and give feed-back to ideas from the non-topic scientists before these ideas are formulated as specific innovation tracks.

Lessons learned

The first phase of the project was completed end of January 2012 with an established and motivated core group consisting of scientist from four Danish Universities. The second phase is in the detailed planning stage presently. So far, lessons learned are that it is surprisingly easy to gather scientists around a new and innovative problem which is outside their basic field of expertise. Furthermore, that it is important to put meat and bone on the problem in question and not revert to too many brainstorming sessions and workshops early on in the process. Finally, that scientists would like to *get into the lab* and do field testing as soon as possible. Again not too many paper based meetings and brainstorming sessions.

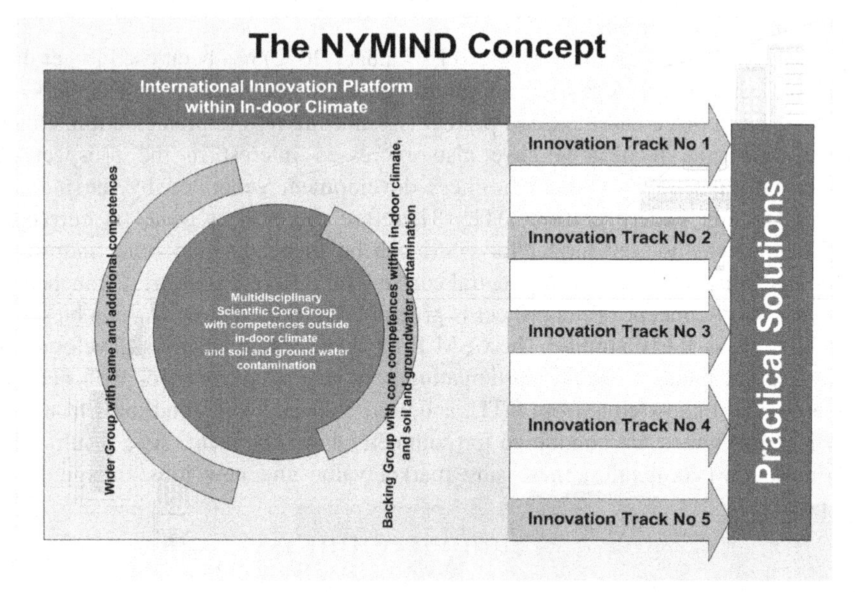

Link:
http://www.regionh.dk/menu/Miljoe/Aktuelle+projekter/Aktuelle+projekter+sider/nyemetodertilsikringafindeklimaet.htm

Case No 4: Municipality of Halsnæs Business Development Project: *CSM 2011 – 2014*

The idea for a Centre of Excellence for Steel Production and Metal Manufacturing (CSM) was fostered by a public authority (Halsnæs Municipality) and a university (The Technical University of Denmark, (DTU)). CSM is organised as an institutionalized Public Private Partnership (PPP) funded by the Danish Government under The Danish Business and Innovation Fund and the Capital Region of Denmark.

The PPP organisation is chosen in order to investigate how a new model for fostering company growth in remote areas of Denmark will work compared to well-known approaches. These more traditional set-ups include public organised counselling services without risk-sharing from the public services and/or private agencies that deliver the service.

The CSM initiative is mainly related to SMEs and there is a great interest from the local steel and metal companies to participate. It takes into account that innovation and increasing productivity occurs when universities, companies and customers interrelate and develop new or substantial improved products or services in unison. It is not a matter of just pushing new products and service to the market. On the contrary, it is a matter of listening to the demands from the market combined with an understanding of future societal needs such as sustainable growth (Chesbrough, 2003).

A new model in Denmark

The first phase of the project (May 2011 - January 2012) has been used to get this complicated structure organised and approved as it is quite new in Denmark. It seems to be in place and the CSM project is venturing into implementation. Other industries in the local area have also expressed interest in the prospect of innovation capacity and future business development generated by the *pull* of knowledge and expertise from DTU. Therefore the project group is currently researching whether the model for CSM can be translated into other industrial areas analysing the effect of one central centre with different industrial branches.

The overall concept of the project is given in the figure below. As can be seen, the principle is fairly simple: The CSM Project processes business development strategies and plans from the participating companies, spices them with needed research based knowledge from DTU, and ensures framework conditions from the Municipality which are conducive for commercial development. As a result, new products, new companies, increasing market value and new jobs emerge from the pot.

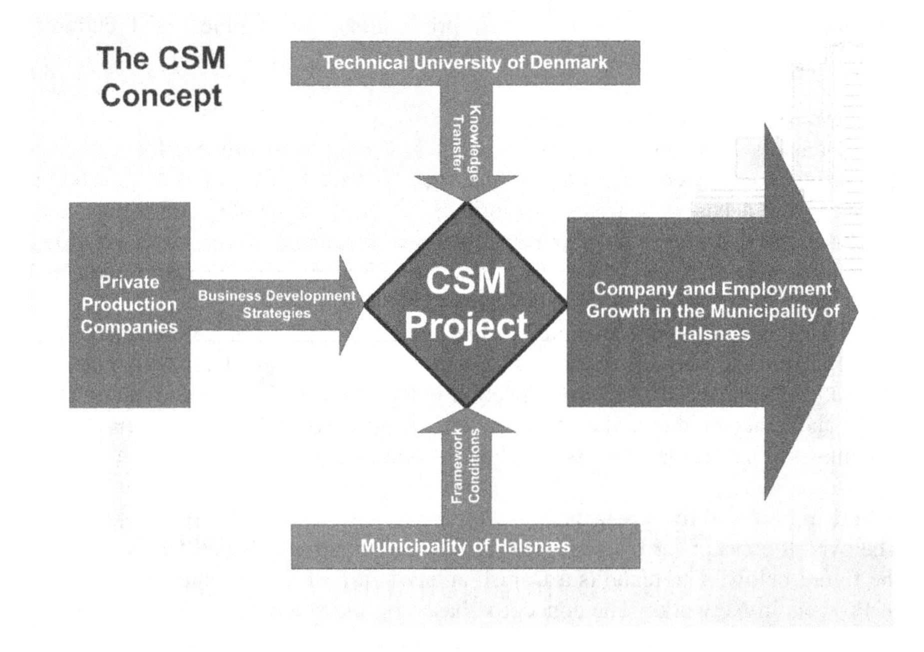

Lessons learned: Private enterprise has warmed to the concept
So far the lessons learned are that the companies located in Halsnæs Municipality have recognised CSM as an open door where public authorities and agencies are helping with connections and networking activities as well as university based knowledge transfer. In this sense CSM is an illustration of the new demand pull approach from companies to universities. The project distinguishes itself from other public-private co-operations in Denmark because the political and administrative system in the municipality takes an active part in the running of the Centre of Excellence for Steel Production and Metal Manufacturing. Another important aspect of the project is the pull-effect the DTU as project-partner will be able to create for the steel and metal companies, through their assistance in the research based product development process.

3 Technical University of Denmark: New Post-graduate Education in Design and Management of Networking Projects

During the last four to five years, the Technical University of Denmark (DTU) has become increasingly aware that globalisation requires special project management tools and skills that are not offered by the traditional project management courses. Special attention has been drawn to the need for methods and skills that enable project managers to design and manage projects in networks which are diverse and complicated with regard to functions, professions, geographies, and cultures (Cicmil et al., 2006; Norman, 2011).

Demand driven education

Based on this DTU in late 2010 launched a development process for a course addressing these needs. To ensure that the envisaged post-graduate education targeted the needs and wishes of private as well as public companies and organisations the whole development process was demand driven based on close interaction with executives from the private as well as the public sector (Hodgson, 2011; Wenger, 2000; Nahapiet & Ghoshal, 1998).

This innovative process resulted in the launch of a post-graduate education in spring 2011 and the first students started in the autumn 2011 with planned graduation in May 2012. As the education is based on fresh research results from DTU and situated there, the students will acquire special skills in relation to planning and managing projects based on the demand driven pull approach.

Practising on real life projects

The overall concept for the post-graduate education named *DELPRON* is given in the figure below. The name is a Danish abbreviation of Design and Management of Projects in Networks. The concept is based on the students working with their present real life projects and not artificial cases. As can be seen it is a precondition that the students have the necessary tools and methods related skills and competences from basic and advanced courses and education in project management.

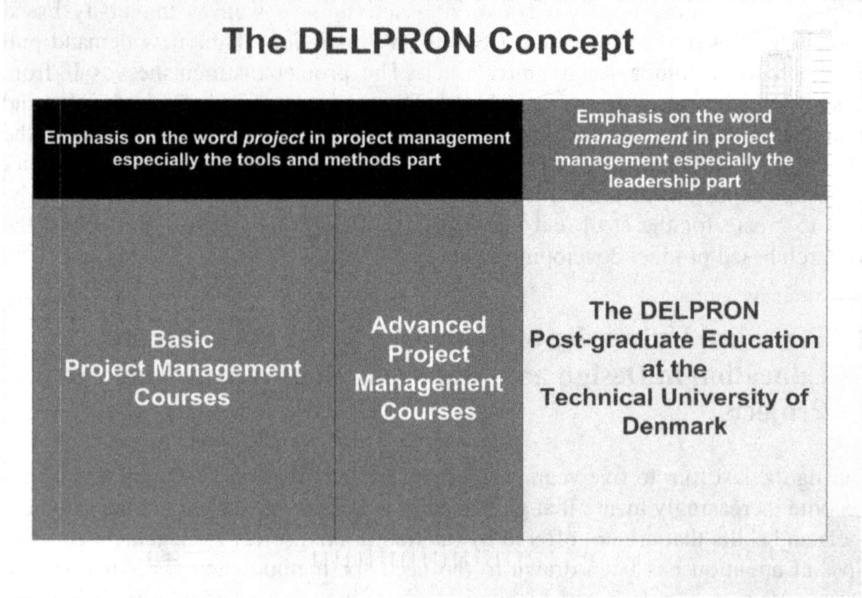

DELPRON focuses on management and leadership in multi-disciplinary, multi-stakeholder, and cultural cross cutting projects implemented in a networking environment. As mentioned before, the education has just started and the preliminary lessons learned are that the students are satisfied with the scope and content of the DELPRON concept. However, students as well as teachers are challenged with the demand driven pull approach as both sides are used to traditional class room education. In addition to this the teachers with their academic results and background have to match the comprehensive real life experience represented by the students in order to create a beneficial knowledge transfer environment. Consequently emphasis is put on bridging this in the present on-going course.

Link: http://transformation.man.dtu.dk/

4 Discussion

Universities produce, disseminate and exchange knowledge in relation to the following three major tracks:

- Scientific papers for peer reviewed journals.
- Candidates for research, for private companies, and for public services.
- Knowledge transfer functions.

The first two should naturally be based on high level research and are mainly push oriented. To some extent there is a pull dimension in relation to candidates based on dialog with the receivers of the candidates. But it is normally not very structured and not very comprehensive as it is a sensitive issue in relation to the independence of universities and the *free research* concept. Consequently there is room for improvement here. However, this is not the topic of this paper.

The focus of this paper is the third track: *knowledge transfer functions* or more to the point *knowledge exchange function.* Universities should focus much more on the demand driven pull function especially in relation to the SMEs. This was revealed by the comprehensive development project ***ViTiS*** conducted at DTU. The knowledge exchange function should basically have the competences and capacity to deliver and exchange knowledge based on needs identified by the SMEs.

Traditionally universities have focused on the push based model as a provider of knowledge not being aware of the special target group that the SMEs represent. In addition to this it seems that the awareness of the importance of the internal barriers for the pull based model is generally not fully developed. Consequently universities should carefully consider their role and functioning as a provider of knowledge especially in relation SMEs. This calls for a top management anchored and well thought through specific strategy in relation to SMEs.

In this connection it is questionable if universities should have a more operational role in relation to SMEs. A role that is not naturally for the universities and which very few have the inbuilt competences to unfold professionally especially concerning business development and not least facilitating the complicated *cultural* meeting between researchers and SME managers. Consequently it is worth considering, and taking into the said strategic development process of the individual university, if a special networking capacity outside universities and SMEs but closely connected to them and with the said competences should be developed. Then universities could focus on their key role as provider of knowledge based on specific demands defined by the said network facility.

Turning towards the SMEs, the demand driven pull based approach requires that especially the SMEs have the insight and capability to:

1. Identify where a transfer/exchange would support innovation and new products.
2. Effective and efficient receive and exchange knowledge.

This is one of the major lessons learned from the ViTiS development project. Combined with the internal barriers within the universities themselves, and the above questionable operational role of the universities, this is the Achilles heel of the demand driven pull based approach. This underpins the need for professional people in the knowledge exchange function related to universities as recommend above. Staff should be professionals in both research and business development. It is a fact that without these in house competencies, the ViTiS project would not have been successful nor would it have had the impact observed.

The City of Knowledge and Urban Creativity Project initially centred its knowledge exchange around the university based Match Function at the main public library. This construction not only confirmed the need for the pull based model. It also emphasised that the university could and would not act as a day to day operator of a professional exchange function. It could achieve more by providing the right knowledge to the right purpose in the right way to the right customer defined through a pull based function as described above. Finally it revealed that – at least in a Danish context – no public authorities or professional commercial organisations will take the full responsibility of establishing a professional and formal facility/facilities for university research based open innovation. They simply do not see this exchange as one of their core obligations. In other words, the networking facility as describe above is presently at least in a Danish context left as an orphan.

The cross municipal climate change mitigation project, ***KLIKOVAND*** has investigated and unfolded successfully the demand driven pull based concept in relation to municipal innovation needs. Furthermore, it is built on professional networking with pure research based development projects. Only three months into the project, one of the major lessons learned is that it requires comprehensive and extensive project management skills to develop and mature such a project. It is expected that the requirements would not soften during project implementation. Consequently there is a need to provide public project managers with skills as described in the newly developed post-graduate education at the Technical University of Denmark (se before and see below).

The ***NYMIND Project*** has revealed that it is fairly easy to engage university researchers in demand driven open innovation processes. The project is concerned with in-door climate problems stemming from contaminated sites or groundwater. Furthermore, it has revealed that especially natural and technical scientists need a fairly firm framework from the onset of their work. Not only in relation to the issue at hand but also the work ahead. Finally, the project shows that the scientists involved prefer more technical based innovation processes as compared to innovation processes which are prone to more creative exercises.

The ***CSM Project*** within steel production and metal manufacturing confirmed the previous lessons learned concerned with open demand driven pull approach and the university as a knowledge provider. The project also found that it is surprisingly easy to create a two-way beneficial knowledge exchange process provided that the right researcher cooperates with the right technical and business staff. As in all walks of life, personal chemistry between the two parties was decisive. The base of the organisation, the PPP (Public – Private - Partnership) model itself, took a long time to formulate, agree upon and mature as it has to deal with several some time contradicting legal, administrative, strategic and political agendas within the partner organisations.

The ***DELPRON*** post-graduate education, addresses some of the competence development needed for running professional knowledge exchange functions, and complicated innovation projects as the ones presented in this paper. DELPRON itself has been developed based on a demand driven open innovation process. The basic concept of using real life projects that the participants are working with in their daily professional life and not artificial cases as background for the education has been well received by the participants which represents a variety of working fields within the public as well as the private sector.

5 Conclusion

As discussed above and illustrated in the figure overleaf, there should be a shift from a pure push based model to a demand driven pull model with the right balance between push and pull, and an overlapping joint area where the exchange is planned, structured and implemented between the *producers* and the *consumers*. This could be organised as an independent networking facility professional in research as well as business development.

As pointed out in this paper it is important to professionalise all three areas in the above figure:

- The strategic as well as operational diagnostic and receiving capacity of especially the SMEs needs to be strongly developed. This should be done in order to professionalise the demand driven pull function based on an understanding of the need of the individual SME for research based knowledge for innovation processes.
- The transfer functions need to be supplemented and professionalised with experienced resource persons. They should have a background in research and business development, and be able to understand and facilitate business

related technology transfer based on the preconditions of the research sector and the commercial sector. Resource persons with this kind of competences will have the capacity and legitimacy to move in and between the circles and especially *occupy* the important shared transfer room and the networking facility.

- As the main producer of research based knowledge the universities should understand their role in the pull based model, and take the necessary strategic, operational and organisational steps in order to further professionalise their capacity within this area. They should naturally continue with developing their dissemination functions but the exchange function needs special focus. In this connection they should carefully consider what should be the proper role and functioning of the universities especially in relation to the more operational phases in the demand driven pull model.
- Last but not least, the SMEs and the universities should further the dialogue in order to improve the understanding and acceptance of their different roles and functions and hopefully by this changing organisational cultures, which could in some cases function as more severe barriers than the professional capacity.

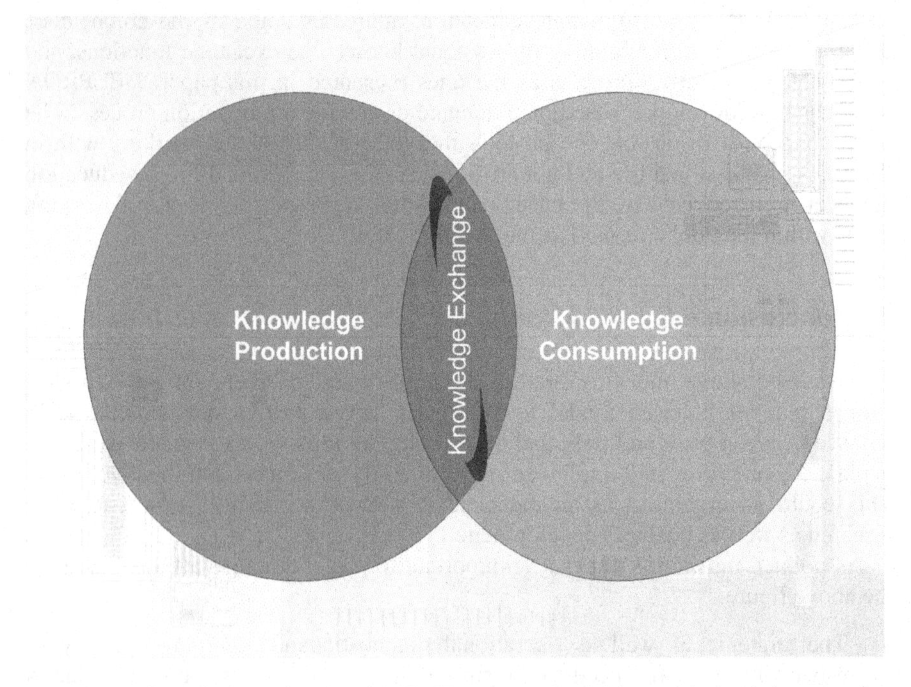

In essence the whole system - the SMEs, the universities and the joint transfer function - needs to be comprehensively and co-ordinated professionalised. And time has come for moving away from the perception that technology transfer is an administrative function to defining a profession called: *Knowledge Exchange Facilitator.* This profession should be developed, educated and capacitated based on proper research within this area.

References

Amternes Videnscenter for Jordforurening. Indeklimapåvirkning fra forurenede grunde. Teknik og Administration nr. 1 (2002) (in Danish)

Chesbrough, H.: Open Innovation. New imperative from creating and profiting from Technology. Harvard Business Press (2003)

Cicmil, Williams, Thomas, Hodgson: Rethinking Project Management: Researching the Actuality of Projects. International Journal of Project Management 24, 675–686 (2006)

DTU ViTiS – Knowledge to Society – debriefing 2008-2012 (November 2011)

EU Commission: White paper - Adapting to climate change: Towards a European framework for action, Com 147 final (2009)

Hodgson: Disciplining the Professional: The Case of Project Management. Journal of Management Studies 39(6), 803–821 (2002)

Nahapiet, Ghoshal: Social capital, intellectual capital, and the organizational advantage. Academy of Management Review 23(2), 242–266 (1998)

Norman: Living with Complexity. MIT Press (2011)

OECD: Economic Aspects of Adaptation to Climate Change (2008)

Rambøll: Kommunernes Investeringsbehov i forbindelse med klimatilpasning og veje (April 2009) (in Danish)

Vandsektorloven: Lov nr. 469 af 12 (June 2009) (in Danish)

Wenger: Communities of Practice and Social Learning Systems. Organization 7(2), 225–246 (2000)

Embedding Creative Processes in the Development of Soft Body Armour: Physiological, Aesthetic, Functionality and Strategic Challenges

Janet Coulter

Research Institute for Art and Design (RIAD), University of Ulster, Belfast, UK

Abstract. The paper describes how creative processes were applied to developing a new product range, to enhance performance, streamline production, reduce costs and open up new markets for body armour solutions. It describes how a company with a proven track record in manufacturing, had lost market share during the first Gulf war due to supply chain issues and had found it difficult to regain share, as the market place had shifted in terms of design. The paper details internal and external challenges in bringing creative processes to the company, to enhance personal protection, improve performance and deliver on aesthetics. Social contexts and physiological issues brought additional challenges. From a business operations perspective, valuing and solving design problems had to be cost effective to manufacture using existing technology. The paper outlines how an academic in fashion and textiles used anthropometrics to research and develop creative solutions for soft body armour. The innovative, new products are now used by security forces globally.

Keywords: Design led, User-centred design, Anthropometrics, Innovation, Knowledge Management, Knowledge Exchange.

1 Introduction

HE institutions in the UK contribute significantly to economic development and underpin a considerable portion of the knowledge economy. The creative industries are growing rapidly and are worth about £0.5 Billion to the Northern Ireland economy alone. The ability to use creativity to generate income has significant advantage within the knowledge economy and partnerships between higher education, government and industry can offer low risk, highly effective solutions in volatile, economic climates.

The company partner is a manufacturer based in Northern Ireland, producing soft body armour solutions. At the commencement of the project, the company had a proven track record in manufacturing but had lost market share during the first Gulf war due to issues with the supply chain.

R.J. Howlett et al. (Eds.): *Innovation through Knowledge Transfer 2012*, SIST 18, pp. 79–89.
DOI: 10.1007/978-3-642-34219-6_8

Having little design experience and found it difficult to regain market share and the key challenge facing the company was how to increase growth and profitability in a dynamic environment, where new competitors were entering the marketplace. It was within this context that a partnership was formed between the company and The University of Ulster.

Investigative research led to a review of the company's business strategy and its approach to design. An analysis and evaluation were presented to management, identifying several key objectives to be addressed through a knowledge transfer project (KTP). KTPs are government sponsored knowledge and technology transfer programmes that enable and support interaction between academia and industry to the benefit of both partners. The key objective is that knowledge and academic expertise is transferred to industry and 'know-how' and industrial expertise is transferred to academia through an industry-based project.

New business strategies were implemented, with design led innovation as the key competitive strength. An amalgamation of several strategies enabled the company to address its challenges from a design management perspective. The business aim was simple, to develop a sustainable business strategy that would regain market share and increase profitability, however the challenges were complex and required a multi-faceted approach. Two case studies are briefly overviewed to demonstrate how some goals were achieved.

2 Rationale for the Research

SMEs are facing strong competition globally and adaption to change provides the key to sustainability and continued success. An internal analysis carried out by the company prior to the project commencing, revealed that creativity and design had been neglected and having neither the internal resources, nor an understanding of design, they approached the University and a collaboration was formed to develop new products, to enable diversification into new markets. The company had managed to trade successfully with a basic product range but wanted to increase market share. Its products, which met all Ministry of Defence specifications, had not previously needed to have visual appeal but clients were now asking for more design alongside functionality, for example **'low-vis'** solutions, when 'soft policing' in communities was required and enhanced, *'high-vis'* solutions for riot situations. With no design capability, the company did not understand how to value design or innovation. The primary challenge was to get the company to recognise, understand and value design and then guide it towards embedding design at the core of all its strategic decision-making.

Companies can often fail to see design as a strategic asset, however research shows that companies which embrace it, tend to be more successful in the market place (Cooper & Press, 2000). Olins (1990) notes that companies must recognise design as a vital resource for it to become part of a management strategy and bringing effectively managed design on board can directly contribute to sustainable growth and commercial success.

The main issues were in relation to improving functionality of design, however aesthetical, physiological and social aspects needed to be considered. Every designed product provides human experiences (Press & Cooper 2005) and consideration of human factors and user-centred design became pivotal in the product development. In applying creative processes several problems were identified and resolved in terms of ergonomics, anthropometrics and aesthetics. The project needed to be driven by in-depth knowledge of how textiles drape, move, react, and contour around a dynamic body, combined with an understanding of visual impact and functionality in clothing. With these factors in mind, it was determined that a fashion designer would offer the best skills set.

3 Setting Objectives

Knowledge management has become increasingly important to companies and Grant (1996) advises that knowledge integration is the primary purpose of any organisation and that knowledge transfer is a key mechanism for knowledge integration (Grant and Baden-Fuller, 1995). An internal review by the Academic revealed that the lack of design knowledge in the company was hampering current operations and this was likely to continue. The company's ad-hoc approach to design would no longer suffice and the review conclusions formed the objectives for the project. The company needed to:

- Streamline production
- Embed an in-house design ethos throughout the company
- Enhance product range to attract a more diverse customer base
- Create Product Data Management systems (PDMS) to manage knowledge and information
- Invest in CAD and CAM systems to develop design efficiently
- Develop a 'Fabric Strategy' to aid Research & Development of new products

3.1 Streamline Production

A key objective was to reduce manufacturing costs and streamline production. An analysis of productivity revealed that the company offered over 100 size variations of its product, which hampered output on the production line and such diversity was not sustainable or cost effective. Anthropometrics were used to undertake a nationwide sizing exercise, measuring forces across the UK, including Group 4 Securicor and Devon and Cornwall Police. The data was analysed and applied to advanced pattern engineering techniques, resulting in a reduction of sizing options by over 50% from 100 to 45 sizes. The exercise was significant as it demonstrated to the company that there were tangible benefits to employing innovative, design techniques. An additional benefit was that excellent relationships and personal service with the existing customer base was maintained.

3.1.1 Embedding Design: A Company Wide Design Ethos

Embedding design throughout the company required a complete culture shift. Von Stamm (2008) cautions that innovation alone is not sufficient contending that the behaviours of people in the organisation need to change. She suggests a *'people based'* approach to ensure successful outcomes and advocates the formation of an innovation team, when innovation requires a different culture and mindset from routine business. This framework was adopted for the project, to realise quick results, which could then be rolled out into the company over a longer period of time.

The organisation had to commit to making changes in its approaches and processes, to ensure the investment in design paid off. Jenkins (2010) describes how organisations are often antagonistic to design, suggesting that existing behaviours and engrained attitudes can potentially *'squeeze the life out of design'* unless addressed. Concurring with Von Stamm, he recommends an entire cultural transformation, concluding that having a designmchampion could be a significant catalyst for cultural change. The KTP Associate (the designer) acted as a conduit to integrate a companywide approach to design, embedding value and expertise to achieve innovative, new products. Thakara (1997) acknowledges that innovation is a major contributor to the ongoing success of any enterprise.

The project initiated the implementation of a design strategy at all levels, as identified by Kelley (2004) and Johnston (2003). Hands (2009) contends that developing design expertise may be especially critical for competitive and growth purposes within global markets; however he acknowl5 edges that it is only part of the picture and design needs to be carefully managed in line with the company's overall strategy. Cooper and Press (2000) concur and recommend blending design with effective management to ensure successful implementation of innovative new products, concluding that fostering creativity should be a specific management goal.

3.1.2 Embedding Design: New Product Development

Innovative design is pivotal to creating new opportunities, nurturing interdisciplinarity and leading to new collaborations, which allow for a holistic approach to radical new product development. The company needed to increase its competitiveness and this would best be achieved through product innovation and diversification into new markets. It can be financially risky to diversify both product and market simultaneously, but at the time this strategy was deemed necessary. Companies can gain competitive advantage during recession through innovative approaches to product development, whilst entering new markets, however they need to be able to increase both efficiency and innovation during a downturn (Geroski and Gregg, 1994 and 1997).

Rhodes and Slater (2009) argue that turbulent times bring with them opportunities as well as threats, a view supported by Deans et al. (2009). The recession had led to several changes in key personnel in the company and its priority needed to be adapting quickly to new challenges. Hartman (2009) suggests that flexibility and rapid response to changing conditions are key objectives during periods of uncertainty. Radical innovation in the company's business model, as described by Leifier et al (2000) was necessary to harness future success. Most new products favour lower-risk incremental innovation, as radical innovation may bring further uncertain dimensions of new technologies and new markets (McDermott and Colarelli O'Connor, 2002); however fundamental changes in approaches to new markets and products were vital for the company's economic survival.

New markets were investigated, including paramedics and with these came added design challenges. Research at the time showed that knife crime had increased significantly, rising by 26.9% between 2006-2007 [1], with frontline health workers being most at risk of close range stabbings. In 2007 there were 1006 physical assaults against ambulance staff in England [1] and many Health authorities continue to report increasing attacks on frontline staff (East of England Ambulance Service reporting a 10% increase in assaults and South Central, reporting a rise of 22% in 2010-11) [2]. This statistical evidence inspired the development of new products to meet the needs of paramedics, such as anti-stab vests, which resist knife penetration in a way that bullet proof vests do not, with high flip collars to counter neck stabbings and attacks from behind. Innovative, user-centred design features were developed such as heat management, back support and systems-orientated designs to ensure compatibility with other on-the-job equipment. Other new product innovations included a 'quick release' system to allow the wearer to swiftly remove armour and themselves from high-risk situations, thus increasing speed and mobility, for example from a burning vehicle, or if the wearer fell into water and needed to off load weight to prevent drowning. Quick release systems also allow medics to attend injured soldiers more easily in the field. This design has continued to be improved in close co-operation with special forces units deployed in conflict zones in Iraq and Afghanistan.

3.1.3 Embedding Design: External Considerations

The company needed its new core values in design to be communicated effectively to its customers and had in its favour its brand name. The company was based in Northern Ireland, whose political history has led it to be associated with high quality, body armour; the 'Northern Ireland' symbolism provided customer confidence and a badge of assurance. However, the new products needed to deliver on performance and create additional value to the core function of protection. Value innovation is based on a deep understanding of the organisation's culture and potential and the needs and desires of the end-users (Fraser, 2010). It is widely accepted that 'value' is a key determinant of customer loyalty and Parasuraman and Grewal (2000) note the impact of technology on the relationships between quality, value and customer loyalty. The project utilised a range of creativity tools, to define the current relationship that the company had with its customers and used the data to project the relationship that it *desired* to have with its customers. The differences between the two datasets, known as the 'Innovation Gap' (Estrin, 2009) were analysed and the results used as a tool for closing the gap to enhance customer loyalty. Mellis (2003) contends that involving customers in the innovation process leads to true partnership, maximising the value to all parties. The company's strong relationship with its customers, led to a bespoke service, accommodating special requests, such as customised vests for officers returning to work after illness, with a colostomy bag or after a mastectomy.

3.1.4 Embedding Design: Internal Considerations

The company's ad-hoc approach to design had meant that decisions were often reactive and extemporised, resulting in acquired knowledge, which was disjointed and difficult to quantify, as the company struggled to determine the intrinsic value of it. This had allowed pockets of 'informal power' to build up. There was some resistance to the ideas of change that the KTP project was proposing, with a perception that this might displace the 'power base'. Resistance to change, first observed by Lewin (1947) is common in organisations (Dent and Galloway Golden, 1999). This posed an unforeseen challenge to the project, which needed to be addressed.

Cohen et al (1990), argue that an organisation's 'absorptive capacity' *(The ability to recognise the value of new external information and apply it to a business model)* depends on individuals and their co-operation on transfers of knowledge across boundaries and sub-units within a company. They continue that 'gatekeepers' (in this instance the KTP Associate) can facilitate the transfer of knowledge across boundaries to increase reception of new ideas, particularly in turbulent environments. Existing company structures needed to be addressed; resistance to change had to be overcome, design needed to be valued and shared between departments and strategies to survive the recession implemented. The introduction of an inhouse design capability provided new focus and *'buy in'* was eventually achieved by involving all departments: technical, quality control, purchasing, shop floor and senior management, facilitating and embedding design at the core of strategic decision-making. All parties worked closely throughout these challenges, with successful outcomes.

Esslinger (2009) stresses that all stakeholders need to understand the fundamental role of creativity in strategic business approaches and presents a useful, three-step, strategic framework comprising: groundwork, creative collaborations and marketing. This strategic framework was applied to the project. As the team gelled, creative collaborations developed and the term 'knowledge transfer' no longer accurately described the project. It became a journey of 'innovation transfer' and 'knowledge exchange'; combining the company's significant expertise in ballistics, with the Academic's and Associate's design knowledge. The process of design was neither linear, nor one-way. It became a two-way process, with a significant culture shift as the change agent. The knowledge acquired was through a shared experience as trust in the partnership was established.

4 CAD Based Systems to Manage of the Design Process

Prior to the project, designs were hand drawn making modifications slow and open to error. Additionally, there was no effective mechanism for managing technical patterns. Lack of communication between departments left product information difficult to control. A product data management (PDM) system was introduced to manage the life cycle of the each product. The PDM was held centrally and could be accessed and tracked by all departments, ensuring transparency, accuracy of information and better workflow processes. New 'Windows' based software for technical pattern management was introduced to create and modify patterns efficiently, enabling the design team to manipulate them to optimise material utilization and reduce material costs. New design software meant designs were produced more efficiently, were visually more attractive to potential new clients and easily understood by pattern cutters. Presentation improved dramatically and this enriched marketing capabilities.

Towards the end of the project, new investors bought over the company and recognising the potential of CAD/CAM based systems, made substantial investment into an automated cutting system to augment speed and precision in production. All four CAD systems were compatible with each other and their integration gave greater control, management and efficiency over the entire design and production process.

5 Smart Fabrics Strategy

Body armour enables humans to function in extreme environments, however it limits heat transfer, impeding activity and performance due to the increased weight and poor moisture management properties. Performance textiles have a high potential for payoff in the fight against terrorism (Jayaraman et al 2005). Improving the textile properties of the vests, alongside ergonomics contributed to overall, enhanced performance. Loose fitting garments allow for better heat transfer (Daanen et al, 2006), but body armour must be close fitting to be functional. A fashion designer's knowledge was critical to the project, as it brought understanding of how materials could be effectively combined to add competitive advantage.

The designer was also able to establish trusted relationships with specialist suppliers and build a performance-fabrics database to augment product development. Atkinson (2003) stresses the importance of building relationships with supply chains to promote innovation in a company. Many suppliers are experts in their field and collaboration early on in the design process can contribute to the finished design and enhance competitive advantage, reducing costs, before they are built into a specification.

6 Research and Development

Research and development feeds design with new technologies, materials and processes (Cooper and Press 2009) and provides a focus and strategic direction. R&D gave rise to two additional research opportunities during the project and although specific details of findings are outside the scope of this paper, the projects have been overviewed.

6.1 Case Study 1: Moisture Management

Investigative research was undertaken to determine if innovative design and performance fabrics could be engineered to improve moisture management, potentially reducing the associated physiological effects. Armoured vests have up to fifteen layers of laminated Kevlar sealed inside a polyurethane cover, causing poor wicking properties. This combined with high external temperatures and/or stressful situations causes the wearer to perspire heavily, potentially leading to dehydration and disorientation, reduced judgment and cognitive performance abilities (Gopinathan, 1988). A six week study at the University of Ulster (Coulter et al 2009), using live subjects and an acclimatisation chamber, tested six prototyped armoured vests using a number controls and variables. Robust statistical applications to the dataset were used to determine the best combination of design and performance fabric. The results were then made available to the company.

6.2 Case Study 2: Moulded Thermoplastic Composites for Female Body Contouring

Stitching darts into multiple layers of Kevlar meant that 30 darts were required to shape each female armoured vest. This labour intensive process gave crude results, reducing aesthetic appeal and functionality. The author's research investigated whether multiple layers of Kevlar could be moulded, three dimensionally over a double curvature in one operation to enhance the ergonomic shape for females. A collaborative experiment between Engineering and Textiles Fashion at the University (Coulter and Archer 2009) was initiated to determine if it was possible to design and manufacture a 3 dimensional multi-layer Kevlar component, using thermoplastics. A body scanner was used to obtain accurate data on female anatomy. This data was used to successfully mould multi-layers or Kevlar onto a female torso shape in one operation, using thermoplastic techniques. The findings were presented to the company.

7 Conclusions

The overall project met with many diverse and unforeseen challenges, which spanned management, communication, technology, production and design. The issues were complex, however through creative approaches, all main KTP objectives were met and challenges largely overcome. The project demonstrates how knowledge exchange has superseded knowledge transfer and how this two-way process can be a rich resource for both universities and industry.

Considerable design and creative knowledge was transferred to the Company, allowing design to be embedded and placed at the core of strategic decision-making. Additional new knowledge was created through interdisciplinary, collaborative research, which facilitated the Company in gaining competitive, commercial advantage. Significantly the outcomes of the project show how design knowledge can contribute to economic growth and demonstrates how important innovation is to the creative industries and the changing economy. The success of the project can be measured through its outcomes. Contracts with many new customers such as: G4S, Devon and Cornwall Police, Gwent Police, Italian Air Force, Trinidad and Tobago Police and Belfast City Council were secured. Additionally a NATO contract, the biggest in the company's history was secured, with a first order placed for 6000 vests. The company continues to trade successfully and uses its design expertise to bring innovative new products to new markets, with recent examples including protection vests for the media in war zones, de-mining suits for aid workers and casual wear with covert, anti-slash protection built in.

Acknowledgements.
Mr. Gary Hayes – Sales Director, at Global Armour Ltd, Lisburn, UK (KTP Company supervisor)
Mrs Orlagh Carmody – Head of Design at Global Armour, Lisburn, UK (KTP Associate)
Dr. Edward Archer, Research Fellow. Engineering Research Institute, University of Ulster
Dr. Gareth Davison, Senior Lecturer in Sport and Exercise, University of Ulster.

References

Atkinson, D.: Supply Chains: Collaborate to Innovate. In: Jolly, A. (ed.) Innovation; Harnessing Creativity for Business Growth. Kogan Page, London (2003)

Cohen, W., Levinthal, D., Damiel, A.: Absorptive Capacity: A New Perspective on Learning and Innovation. ASQ 35, 128–152 (1990), http://faculty.babson.edu/krollag/org_site/org.../cohen_abcap.htm (last accessed January 10, 2012)

Cooper, R., Press, M.: The Design Agenda; A guide to Successful Design Management. John Wiley & Sons, UK (2000)

Coulter, J., Davison, G., McCLean, C., Carmody, O.: 'What if?' Research and Innovation by Design. In: Proceedings of the Senior Staff Conference, University of Ulster (2009)

Coulter, J., Archer, E.: "What if?' Research and Innovation by Design. In: Proceedings of the Senior Staff Conference, University of Ulster (2009)

Daanen, H.A.M., Reffeltrath, P.A., Koerhuis, C.L.: Ergonomics of Protective Clothing. In: Jayaraman, S., et al. (eds.) The Netherlands in Intelligent Textiles for Personal Protection and Safety. IOS Press, Amsterdam (2000)

Deans, G., Kansal, C., Mehltretter, S.: Making a Key Decision in a Downturn: Go on the Offensive or be Defensive?. Ivey Business Journal Online (January/February 2009)

Dent, E.B., Galloway Goldberg, S.: Challenging "Resistance to Change". Journal of Applied Behavioural Science 35, 25 (1999), doi:10.1177/0021886399351003

Esslinger, H.: A fine Line: How business strategies are shaping the future of business. John Wiley & Sons, San Francisco (2009)

Estrin, J.: Closing the Innovation Gap: Reigniting the Spark of Creativity in a Global Economy. Mc-Graw Hill (2009)

Fraser, H.: Design Business: New Models for Success. In: Lockwood, T. (ed.) Design Thinking. Allworth Press (2010)

Geroski, P.A., Gregg, P.: Corporate Restructuring in the UK during the Recession. Business Strategy Review 5(2), 1–19 (1994)

Geroski, P.A., Gregg, P.: Coping with Recession. UK Company Performance in. Adversity. University Press, Cambridge (1997)

Gopinathan, P.M., Pichan, G., Sharma, V.M.: Role of dehydration in heat stress-induced variations in mental performance. Arch. Environ. Health 43, 15–17 (1988)

Grant, R.M.: Prospering in dynamically competitive environments, Organisational capability as knowledge integration. Organisation Science 7(4), 375–387 (1996)

Grant, R., Baden-Fuller, C.: A knowledge-based theory of inter-firm collaboration. Academy of Management Best Paper Proceedings, 17–21 (1995)

Hartman, N.: Sure ways to tackle uncertainty in tough times. Financial Times, Managing in a Downturn Series (February 5, 2009), http://www.ft.com/reports/managingdownturn (last accessed December 15, 2011)

Jayaraman, S., Kiekens, P., Grancaric, A.M.: Heat Strain and Fit, TNO Defense, Security & Safety. In: Jayaraman, S., et al. (eds.) The Netherlands in Intelligent Textiles for Personal Protection and Safety. IOS Press, Amsterdam (2006)

Jenkins, J.: Creating the right environment for Design. In: Lockwood, T. (ed.) Design Thinking. Allworth Press (2010)

Johnston, R.E.: The power of Strategy Innovation; A new way of linking creativity and strategic planning to discover great business opportunities. Amacom (2003)

Jolly, A.: Innovation: Harnessing Creativity for Business Growth. Kogan Page, London (2003)

Kelley, T., Littman, J.: The Ten Faces of Innovation; Strategies for heightening Creativity. Profile Books (2004)

Kitching, J., Blackburn, R., Smallbone, D., Dixon, S.: Business Strategies and Performance during difficult economic times. Department of Business Innovation and Skills. Small Business Research Centre, Kingston University (2009) (January 04, 2010)

Kotter, J.P.: Leading Change: Why Transformation efforts fail. Harvard Business Review 73(2), 59–67 (1995)

Leifer, R., McDermott, C.M., Colarelli O'Connor, G., Peters, L.S., Rice, M.P., Veryzer, R.W.: Radical Innovation: How Mature Companies Can Outsmart Upstarts. Harvard Business School Press, Boston (2000)

Lewin, K.: Frontiers in Group Dynamics. I. Concept method and reality in social sciences: social equilibria and social change. Human Relations 1, 5–41 (1947)

Lieberman, H.R.: Hydration and Cognition: A Critical Review and Recommendations for Future Research. J. Am. Coll. Nutr. 26(suppl. 5), 555S–561S (2007)

Lorange, P.: Optimists have a bright future. Financial Times. Managing in a Downturn Series (January 29, 2009), http://www.ft.com/reports/managingdownturn (last accessed December 14, 2011)

McDermott, C.M., Colorelli O'Connor, C.: Managing radical innovation: an overview of emergent strategy issues. Journal of Product Innovation Management 19(6), 424–438 (2002)

Mellis, W.: Partnering with Customers. In: Jolly, A. (ed.) Innovation; Harnessing Creativity for Business Growth. Kogan Page, London (2003)

Olins, W.: The Wolff Olins guide to corporate identity. The Design Council, London 12 (1990)

Parasuraman, D., Grewal, D.: The impact of technology on the Quality-Value-Loyalty-Chain, A research Agenda. Journal of the Academy of Marketing Science 28(1), 168–174 (2009)

Press, M., Cooper, R.: The Design Experience, The Role of Design and Designers in the Twenty First Century. Ashgate Publishing Limited, Hampshire (2005)

Rixon, D.: Strategic Commitment- The Foundation Stone for Innovation. In: Jolly, A. (ed.) Innovation; Harnessing Creativity for Business Growth. Kogan Page, London (2003)

Thakara, J.: Winners! How today's successful Companies innovate by Design. Gower, London (1997)

Rhodes, D., Slater, D.: Seize Advantage in a Downturn. Harvard Business Review 87(2), 50–58 (2009)
Vinding, A.L.: Human Resources; Absorptive Capacity and Innovative Performance. Research on Technological Innovation and Management Policy 8, 155–178 (2004)
Von Stamm, B.: Managing Innovation, Design and Creativity, 2nd edn. John Wiley and Sons Ltd., Chichester (2008)

[1] http://www.ppss-group.com/blog/when-will-ambulance-staff-and-paramedics-be-issued-withbodyarmour/ (last accessed December 15, 2011)
[2] http://news.bbc.co.uk/1/hi/uk/7088341.stm (last accessed December 15, 2011)

The Alchemy Exchange - Turning Student Consultancy Opportunities into a Good Student Experience

Felicity Mendoza and Jonathan Gorst

Sheffield Hallam University

Abstract. This paper describes the journey undertaken by Sheffield Business School, in implementing a student consultancy unit as an extra-curricular activity and the positive and negative issues that it has faced. It is a case study which will take the reader from conception through to a functioning student consultancy service. The Alchemy Exchange has established itself as part of the overall student offering, alongside course based consultancy while making itself part of the Business School's business engagement offer. The case study provides an insight into how an extra curricular activity can be used as a way to provide a cost effective consultancy service to business, while providing students with a real life learning experience for which they are paid, supervised by an academic and transferring knowledge.

Keywords: Student Consultancy, Extra Curricular Activity.

1 Introduction

This case study looks at the work of The Alchemy Exchange (TAE), a student consultancy unit based within Sheffield Business School at Sheffield Hallam University. It looks at how the original concept was developed into a working unit and the challenges it faced in becoming established to provide real life experiences to students.

2 Background

Originally named the Student Consultancy and Research Unit, The Alchemy Exchange was set up in 2009 as a mechanism for raising the profile of Sheffield Hallam University's newly rebranded Sheffield Business School (SBS) within the local business community (Damodaran, L. 2009). As one of a suite of Knowledge Exchange activities which included contract research, Knowledge Transfer Partnerships (KTPs) and Corporate Professional Development (CPD) (Prince

R.J. Howlett et al. (Eds.): *Innovation through Knowledge Transfer 2012*, SIST 18, pp. 91–99.
DOI: 10.1007/978-3-642-34219-6_9

2007), TAE was tasked with generating income and developing new partnerships for the Business School.

In 2009, the choice for external organisations wishing to engage the Business School's consultancy services was between academic consultancy - at prices which often excluded smaller businesses and companies affected by the economic downturn - and free course-based consultancy projects, which could only happen at fixed points in the year. The quality of the latter could not be guaranteed, however, positive feedback from the most successful course-based projects suggested that there was potential to develop a commercial consultancy unit staffed by students, supervised by academics and managed by project facilitators. Client feedback from these free projects indicated a willingness amongst some external organisations to pay for student consultancy providing the quality of the output could be guaranteed. Internally it was felt that without the seasonal time constraints of the course-based projects and with the right students and academic supervision, consultancy projects could be delivered to a professional standard.

3 The Pilot

A pilot was carried out during spring 2009 in which 3 projects of differing values, £1,000, £4,000 and £12,000, were undertaken and it was concluded that the optimum project size was £4,000, which equated to approximately 10 student days with approximately 2-3 days of academic supervision. Whilst all 3 projects were delivered successfully, the importance of a rigorous process to recruit reliable students was highlighted when one student dropped out leaving the Academic Supervisor and PM (PM) to complete the work at short notice.

By late spring 2009 it was felt that there was sufficient potential to create the new role of PM whose task would be to develop the pilot into an income generating consultancy unit. The unit was renamed The Alchemy Exchange (TAE), a name which was chosen to enable flexibility as it grew.

4 The Administration

Initially, TAE was assigned one full time manager, supported by a 0.5 Full Time Equivalent (FTE) placement student. The PM liaised with the Finance department to develop a robust costing model and a daily charge rate was developed which would include associate, academic and project management. A terms of reference template was developed which, when paired with the University's standard terms and conditions for consultancy, acted as a binding contract with the client. It was agreed with Human Resources that students assigned to projects were employed on casual contracts under the role title of TAE Associate and paid £8 per hour.

5 Marketing

Long lead times for the design and production of web pages, flyers and stands meant that the TAE staff had to get the message out by other means such as attending welcome lectures for selected courses, using the University's online vacancy advertising service and attending recruitment fairs.

B2B marketing activities included the development of web pages, flyers and stands and key University personnel such as Knowledge Transfer Champions and Business Development Managers were briefed on TAE's offer to business. Existing contacts were informed via newsletters and TAE was represented at several internal and external networking events.

6 The Students

The opportunity to work with TAE was open to students from all courses and stages of study from across the University, plus recent graduates, however, the majority of successful applicants came from the Business School's business and marketing courses. As the university with the largest number of students on sandwich courses in the UK, TAE was particularly attractive to undergraduate students returning for their final year after their industrial placement. With career success in mind, this cohort sought to maximise the return on their investment in their education by gaining relevant work experience and enhancing their CV ready for the graduate labour market (Lehmann, 2009). Many were also wary of inflexible and demanding part- time work which could impact negatively on their final year studies by obliging them to over commit whilst failing to develop their skills (Clegg et. al. 2010).

Another prominent group included recently graduated international Master's students on Post Study Work Visas. A number of graduates from this group chose to stay in the city once their course had finished but found it difficult to obtain full time graduate level employment. TAE was therefore attractive to them because it offered the opportunity to demonstrate to potential employers their ability to operate on a professional basis outside their country of origin.

7 The Recruitment and Selection Process

One of the key concerns highlighted by the pilot was the risk that students might jeopardise the success of a project and, with it, the reputation of the University by being unprofessional or unreliable. Therefore it was important to design a recruitment and selection process which would test not only the skills and abilities of the candidates but their commitment and motivation as well. In order to mirror the kind of selection activities typically found at graduate level, students were invited to submit an application form, from which a shortlist was drawn up and a selected number were interviewed.

Successful candidates were invited to become TAE Associates and join a 'Talent Pool' of Associates who had access to project opportunities. The Talent Pool was developed in order to ensure responsiveness and avoid delays when a project was being scoped.

8 The Academics

During the pilot of TAE, the academic supervision was undertaken by a principal lecturer from the Marketing subject group. With many years experience of working as a consultant in the UK and Europe in conjunction with the University as well as within the private sector, the academic was well placed to guide the initiative forward.

Internally, TAE was positioned as a knowledge exchange activity, a 'softer collaborative experience' (Ramos-Vielba & Fernández-Esquinas, 2011), which allowed early career teaching academics the opportunity to experience knowledge exchange in a low risk environment with considerably less time commitment than a KTP.

However, the structure and practices of the University (Jacobson et. al., 2005) as well as the priority of teaching and individual achievement over organisational objectives, (Seonghee & Boryung, 2008) meant that interest amongst academics was low which resulted in the original Marketing academic taking on the supervision of 4 of the 6 early projects. There was also nervousness amongst academics regarding the ability to consistently deliver to a commercial standard given that students would be employed to carry out the bulk of the work (Neagle et. al. 2010).

As TAE's reputation grew and the PM's knowledge of individual academics' research and personal interests increased, more academics were attracted to join TAE.

As other academics, new to business engagement and consultancy, joined, the PM was presented with a new range of challenges. This group of academics, whose skills were based on in-depth research of a specific discipline, had to adjust their enthusiasm to meet the constraints of commercial consultancy. The PM was tasked with turning research ideas into cost-effective ways of meeting the clients' requirements within the budget and timescales available (Powell, 2007).

Feedback gathered from participant academics illustrated how they benefited from their involvement with TAE. These academics were able to use examples from their TAE experiences in the classroom and case studies were developed to enable students to work on simulated projects as part of their course.

Traditional knowledge exchange activities have been carried out between universities and industry. The TAE model and the available funding, however, encouraged charities and social enterprises to engage with the University and attracted a different type of academic to knowledge exchange (Reichenfeld, 2011).

9 The Clients

The early stages of TAE coincided with the emergence of regional funding sponsored 'Innovation Vouchers' for small and medium enterprises. The vouchers, worth £3,000, were positioned as a contribution towards consultancy services and gave recipients the opportunity to engage with universities as well as other commercial agencies. The nature of the funding had a significant impact on TAE's client profile and upon its development as a research and consultancy centre.

The £3,000 funding was available to SMEs in the region during the first year of TAE's operation. Despite the funding being intended as a contribution towards the cost of consultancy services, the majority of clients were not prepared to spend over the funding allowance. This led to projects being designed to fit the £3,000 model rather than to address the business issue in a holistic sense. In addition, many clients, due to lack of experience in engaging consultancy services, had unrealistic expectations about the level of detail that could be achieved within their budget and on more than one occasion pushed for additional work to be done over and above the specification.

In some cases there was a mismatch between the original client brief submitted to the funding body and the client's requirements which emerged during the first scoping meeting. It became apparent that the client's funding application was worded to match the funding body's eligibility criteria although the decision to access consultancy was promoted by a problem or change beyond the normal scope of the client's capacity or resources (Jacobson et.al., 2005).

10 Project Management

Once chosen as a supplier, the PM, academic and client scope out the project in detail. A terms of reference document including the background to the project, its aims and objectives, the proposed research activities, timescales, costs and terms of business are then issued to the client.

Once the opportunity had been agreed with the client the specific skills requirements, tasks and timescales were sent to the Talent Pool. Associates in the pool were asked to apply for the position by submitting an expression of interest and confirming their availability. First time associates were required to attend an induction which included briefings on safety for lone workers, equality and diversity, data protection and timesheet management. Associates were then briefed by the PM and academic supervisor about the project itself.

The PM arranged, attended and minuted regular progress meetings between the academic supervisor and the associates, managed the relationship with the client and ensured that actions were completed within internally agreed deadlines.

Nearing the appointed end of the project, the PM liaised with the team and the client to set up a meeting for the presentation of the findings. The project findings were delivered by the associates as a verbal presentation backed up by a full report.

11 Review of Years 1 & 2

After 1 year of operation TAE had carried out 24 projects with an average cost of £3000, reflective of the funding available. The potential of TAE as a resource was recognised within the University and out of the 24 projects completed in Year 1, 5 were carried out for internal clients whilst 17 of the remaining 19 external clients had received Regional Development Agency (RDA) funding.

However, at the end of the first year of operations, regional funding for SMEs came to an end. As the majority of clients had been start ups, micro businesses and not for profit organisations their budget for commissioning future projects was very limited. In fact, none of the funded projects led to self-financed interactions although some did lead to free course-based projects.

In Year 2, TAE completed 20 projects with an average cost of £1,500. Of the 20 projects, 3 were funded, 9 were paid for by the external organisations and 8 were internal.

Due to the lack of available funding it was necessary to offer a reduced cost option in order to continue to attract external organisations. Therefore an alternative TAE costing model was developed so that, in addition to the original model of a bespoke project team consisting of a PM, an academic supervisor and associate(s), potential clients were also given the option to act as the supervisor, overseeing the associates' work with the support of a PM. This had the effect of increasing the number of paid, non funded opportunities in Year 2 but reduced TAE's overall income by approximately 50%. In addition, a number of clients who opted for this model did not have the requisite skills or knowledge to supervise the project, resulting in increased pressure on the PM to provide academic supervision that had not been built into the costing at the project scoping stage.

It also became apparent during Year 2 that some clients were unable, either due to resource limitations or lack of expertise, to implement the findings or recommendations that the project had produced. Costs and other practicalities aside, ideally assisting with implementation and measuring impact should have been built in to the remit of the project (Smith and Paton 2011).

A SWOT analysis of TAE was carried out at the end of Year 2 to identify its key strengths, weaknesses, opportunities and threats:

SWOT – The Alchemy Exchange	
Key strengths	**Key weaknesses**
• Enhancing the student employability and business engagement agenda • Providing academics with links to industry, case studies and staff development opportunities • Cost effective service for external partners	• In a recession, many businesses may not be prepared to outlay costs for consultancy, especially when business support was previously funded by business link • Marketing and business engagement strategy - difficult to articulate offering • Culture / bureaucracy of delivering commercial projects within an academic environment
Key opportunities	**Key threats**
• Integration of UG and PG students across courses • Improving our offer to both students and businesses • Projects can lead to ongoing relationships and other opportunities such as placements, guest speakers and course-based projects	• Cuts in access to funding for SMEs • Reluctance of businesses to spend money and /or time due to economic climate • Availability of students to who can afford to work in an ad hoc, flexible way

Source: Adapted from R. Keeton's Business Engagement Programme final report submission 2011

12 Feedback

As a result of the feedback gathered from all participants at the end of each project, a number of areas of best practice were identified. Projects tended to be more successful when:

- achievable and detailed project plans were developed collaboratively between the PM, academic and client
- realistic methodologies were employed e.g. primary research was only undertaken if the source could be identified in advance
- associates were matched to projects according to their skills, experience, interests and availability over the period required
- the academic supervisor, associate and client were committed to the project and understood their individual roles
- team work and communication were facilitated by the PM. For the majority of projects weekly meetings were used to monitor progress, review findings and delegate further actions.

Feedback collected from associates confirmed the value of TAE as a means of enhancing the student experience and developing employability skills.

13 Comparable Initiatives in Other Universities

During the pilot stage a visit was made to the University of Hertfordshire's Graduate Consulting Unit. Based within their Centre for Entrepreneurial Development, the Graduate Consulting Unit had been running for approximately 3 years at the time of the TAE pilot and their pricing model was influential in developing TAE's own prices. At that time, the Graduate Consulting Unit's management were considering branching out as an independent enterprise.

Typically, however, initiatives developed by other universities tend to focus on IT consultancy. Leeds Source-IT and Genesys Solutions, from Leeds University and The University of Sheffield respectively, employ students specifically from computing and software engineering to work on commercial projects for external clients. According to Neagle (p.267, 2010), the range of skills originally offered to clients were too broad and this led to management issues, therefore by narrowing the skills set on offer they maximised on their strengths and were able to duplicate their successes.

14 Challenges and Lessons Learned

The Alchemy Exchange is an ambitious project which was open to students and graduates from all levels and courses across the University. The initial targets (24 projects in Year 1) and the funding available for local businesses meant that, like Leeds Source-IT (Neagle, 2010), the offer to clients was too broad. In future, TAE will be limited to students within the Business School and will only take on projects which are relevant to Business School courses. By focusing on market research and data gathering TAE will specialise in carrying out feasibility studies, scoping out new products, services and markets and providing content for business plans.

Although the first year targets were met, it was at the expense of developing a business development strategy to source projects once regional funding came to an end. An internal restructure saw the formation of a Business Engagement Team which was primarily tasked with securing high value, corporate contracts and therefore, low value, small scale consultancy projects tended to be overlooked. In addition, as TAE was repositioned as an employability initiative, there were further challenges for business facing colleagues to see its relevance to their remit. As Wolfenden (1995) points out, a coordinated strategy for developing external business relationships is challenging given the issues of 'ownership and communication'.

As one of a number of employability initiatives within the University, it was challenging to articulate TAE's offer both internally, to staff and students, and externally. The aim for the future is to bring all employability initiatives under one coordinating hub which will act as a single point interface between students, business engagement staff, academics and businesses as well as providing coherence and support that will develop and enhance corporate links (Keeton 2012).

15 Conclusion

In its initial stages TAE was tasked with generating income and developing new partnerships for the Business School, however, limited ability to generate significant profit or major corporate leads resulted in its re-positioning. The renewed emphasis on the student experience in the University's Annual Operating Plan, coupled with positive feedback from associates in the Talent Pool allowed TAE to change its focus and develop its strengths as an employability initiative. A change in emphasis on income generation and financial subsidy from the Business School facilitated the development of new employability initiatives under the TAE umbrella including free projects for non profit organisations and internships.

References

Clegg, S., Stevenson, J., Willott, J.: Staff conceptions of curricular and extracurricular activities in higher education. Higher Education 59(5), 615–626 (2010), doi:10.1007/s10734-009-9269-y

Damodaran, L.: The Graduate Consulting and Research Unit Feasibility study (phase 1). Sheffield Hallam University internal document (2009)

Jacobson, N., Butterill, D., Goering, P.: Consulting as a Strategy for Knowledge Transfer. The Milbank Quarterly 83(2), 299–321 (2005)

Keeton, R.: Business Engagement Programme. Final Report (2011)

Lehmann, W.: University as vocational education: working - class students expectations for university. British Journal of Sociology of Education 30(2) (2009)

Keeton, R.: The Employability Hub 2012. Internal Sheffield Hallam University paper (2012)

Powell, J.: Creative universities and their creative city-regions. Industry & Higher Education (October 2007)

Prince, C.: Strategies for developing third stream activity in new universi-ty business schools. Journal of European Industrial Training 31(9), 742–757 (2007)

Neagle, R., Marshall, A., Boyle, R.: Skills and Knowledge for Hire: Leeds Source-IT. In: ITiCSE 2010, Bilkent, Ankara, Turkey, June 26-30. ACM (2010), doi:978-1-60558-820-9/10/06

Ramos-Vielba, I., Fernández-Esquinas, M.: Beneath the tip of the ice-berg: exploring the multiple forms of university–industry linkages. The International Journal of Higher Education Research (December 2, 2011)

Reichenfeld, L.: The Barriers to Academic Engagement with Enterprise: A Social Scientist's Perspective. In: Howlett, R.J. (ed.) Innovation through Knowledge Transfer 2010. SIST, vol. 9, pp. 163–176. Springer, Heidelberg (2011)

Seonghee, K., Boryung, J.: An analysis of faculty perceptions: Attitudes toward knowledge sharing and collaboration in an academic institution. Library & Information Science Research 30(4), 282–290 (2008)

Smith, A.M.J., Paton, R.A.: Delivering enterprise: A collaborative international approach to the development, implementation and assessment of entrepreneurship. International Journal of entrepreneurial Behaviour & Research 17(1), 104–118 (2011)

Wolfenden, R.: Experienced of Enterprise in Higher Education within two research-led universities. Education & Training 37(9), 15–19 (1995)

The Use of Accompanying Funding Offers When Starting Knowledge Transfer Projects: The Case of National Innovation Agents in Denmark

Erik W. Hallgren

Teknologisk Institut, Gregersensvej 1, 2630 Taastrup, Denmark
ewh@teknologisk.dk

1 Principle Topic

A recent analysis demonstrates that a huge undeveloped innovation potential exists in many small and medium-sized enterprises (SMEs) in Denmark [1]. The SMEs do not, in many cases, have the necessary knowledge capacity in-house, and they lack the knowledge about how to initiate and maintain innovation linkages externally. Furthermore, many SMEs do not focus enough on the long-term and strategic development perspective because the short-term tasks often occupy the management full-time, especially given the current financial situation [1]. Thus, they appear differently in very small (<10) and larger (>10) companies, as shown in this paper / presentation. This presentation represents the author's findings regarding whether company size is important for accepting a knowledge transfer project in situations where offered funding requires certain participation from the SME.

During the period from 2007 to 2012, an innovation agent program[1] was supported financially by The Danish Agency for Science, Technology and Innovation as one initiative to cope with the aforementioned challenge. The SME voluntarily enters into cooperation with the Regional Innovation Agents, where the former is given guidance on their innovation plans and projects. One of the main tasks of the innovation agent program is to establish connections and knowledge transfers between Danish SMEs and knowledge resources (universities, networks, research programs, private consultants, etc.). To overcome some of the barriers to knowledge transfers, the program also assists the SME in obtaining funding for the knowledge transfer projects.

One program is the "videnkupon" (knowledge coupon) program, supporting the SME with up to 13,500€ to invest in knowledge; the SME then has to raise (and use) one and a half times that amount of money to cover its own expenses.

[1] Please see www.dti.dk/services/31424?cms.query=innovation+agent for more information.

R.J. Howlett et al. (Eds.): *Innovation through Knowledge Transfer 2012*, SIST 18, pp. 101–102.
DOI: 10.1007/978-3-642-34219-6_10

2 Method

As the author is part of the program and trained in action research [2], the research question was formulated through his work with companies and then verified by analysing the database. Up until now, more than 1000 SMEs have participated in the program, providing data for analyses that show how well advice and contacts for knowledge transfer partners have been received compared to company size, economy, branch, etc. More than 40 agents are participating in the program and 5 small-scale qualitative interviews have also been conducted with agents to verify the findings.

3 Results and Implications

This short paper presents how very small companies compared to larger companies more often accept knowledge transfer projects when the project is funded by a "viden kupon". The larger companies usually hesitate to accept the funding offer with the accompanying demand on the use of their internal resources. Conversely, no difference was found between the two groups of companies in their interest for purely meeting new knowledge partners. This approach gives the larger companies an opportunity to be more selective in choosing the start time and how many resources to spend on a project. However, it also decreases the amount of knowledge transfer projects that are started, especially for larger companies.

This finding should be taken into consideration when planning knowledge transfer programs with included funding schemes in the future.

References

1. Danish Agency for Science Technology and Innovation (DASTI). Innovation Agents: New roads to innovation in small and medium-sized enterprises. The Danish Agency for Science, Technology and Innovation. Copenhagen, Denmark (2009)
2. Hallgren, E.W.: Employee-driven innovation: A case of implementing high-involvement innovation. Academic dissertation. Ph.D. thesis (2009) ISBN: 978-87-90855-16-1

How to Facilitate Knowledge Collaboration – Developing Next Practice

Elinor Bæk Thomsen[1] and Trine Lumbye[2]

[1] Central Denmark Region, Innovation and Business Development, Denmark
[2] AU Centre for Entrepreneurship and Innovation, Aarhus University, Denmark

Abstract. This paper describes developments on new methodologies to facilitate the good and stimulating knowledge collaboration between business and researcher. This work in progress is focused on studying and describing three levels in the knowledge collaboration: *Information*, *knowledge* and *competences*.

1 Introduction

'Shortcut to new Knowledge' is a three year project. The general objective is to further innovation in SME's through knowledge collaboration. In the project we match small and medium sized enterprises (SME's) with researchers from Danish or foreign universities and furthermore facilitate cooperation and collect learning from the processes. The starting point for a collaboration between enterprise and researcher will be a specific innovation project and knowledge demand as formulated by the enterprise. The researcher will contribute with open and public knowledge, tools and methods without any IPR restrictions.

One of the aims with 'Shortcut to new Knowledge' is to develop next practice within facilitating knowledge collaboration. Therefore the project is focusing on experimenting with different methods and approaches for supporting the good and stimulating knowledge collaboration. Not only with respect to the success of the innovation project, but more specifically on the fact that the owners, leaders and employee's in the enterprise obtain new competences and opportunities for action and are stimulated to use these.

2 Approach

We use the term "knowledge collaboration" instead of "knowledge transfer" in order to emphasis the to-way proces. Both researcher and company people have knowledge and it is the combination that enables further development and innovation. We distinguish between three levels of knowledge collaboration: Information, knowledge and competence. When the researcher is presenting his research based knowledge to the company, we will argue that the knowledge transfer has not been successful until the individual employee or/leader is capable of conceptualizing the information. Furthermore if the employee is able to convert

R.J. Howlett et al. (Eds.): *Innovation through Knowledge Transfer 2012*, SIST 18, pp. 103–104.
DOI: 10.1007/978-3-642-34219-6_11

knowledge to experienced positively changed behavior, knowledge has been transformed into competence.

We are mainly interested in the *proces*, from information to knowledge and from knowledge to competence, and we are interested in exploring which methods of facilitation that can support these processes.

Our practical approach to explore and describe this process is conducted in two levels:

1. Experiments: Using different methods of matchmaking and facilitation. Investigating how they affect the knowledge collaboration and innovation in the companies.
2. Describing knowledge: We select a number of knowledge employee's and conduct participation studies on meetings and workshops between enterprise and researchers followed by qualitative interviews with employee's. In this investigation we are mainly interested in the questions: How will employees (and leaders) describe and use the knowledge that is achieved in the knowledge collaboration? And how will the achieved knowledge effect the way the enterprise work with and understand innovation?

3 Preliminary Results

Based on results from the initial phase of knowledge collaborations, we have learned that the facilitation of the introductory matching of expectations between researcher and enterprise is crucial. We have achieved good results combining project management from the company or private consultants with facilitation by university staff. It is important that the different objectives for the participants are articulated up front, and that the diversity is appreciated and respected. We have learned, that researcher are motivated by the possibility to give research a reality check. Even if confidentiality hinders them to use the findings directly in their research. And we have learned that private companies are highly motivated by the chance to develop their business models

The project is still young and during the next few years we will continue to experiment and learn about knowledge collaboration. Topics will emerge through the project. But right now we want to know more about the role of academic and business language, and the influence of personal relations.

The On-Campus Innovation Factory - Boosting University Research Based Innovation

Lars Hein

IPU, Denmark
LH@IPU.DK

Content & main issues

The most powerful and concentrated sources of technology know-how and innovation lies with our technical universities. Making this source of know-how and innovation operational for private and public companies holds a tremendous potential for development and growth. However, most reports on the cooperation between university and industry seem only to document the many problems and pitfalls experienced when the two very different cultures meet and interact. And most analysis on the actual volume of cooperation between university and industry indicates that we fall very much short of the potential that could be reached.

The current motivation of the technical universities to provide innovation services to industry is balanced by two opposing forces:

One force comes from the general acknowledgement from top level management that the university should be industry-oriented (though not industry-driven), and thus its researchers should engage with industrialists for the mutual benefit of researcher and practitioner. Another, and opposing, force comes from the increased use of goals and corresponding metrics for research performance by which the researchers are evaluated. Success and prosperity in the research community depends on sustained high scores in terms of number of journal papers, citations, monographs, etc., and is not perceived to depend on interaction with industry. Therefore, solutions that allow university staff to maintain close contact and deep interaction with industry, while preserving their full capacity for research purposes, is a valuable asset for a university.

This contribution reports on the setup, organisation, and innovative results of such a solution at the Technical University of Denmark (DTU), based on the concept of an on- campus innovation provider with a powerful, permanent staff, acting as a agent to boost the university/industry relationship. The presentation is supplemented by industry cases and examples.

References

[1] Paisley, E., Pool, S.: Getting Good Technology out of the Lab and into the Marketplace. Small Business Innovation (January 2012)

[2] Holly, K.: The Full Potential of University Research. Science Progress (June 2010)

R.J. Howlett et al. (Eds.): *Innovation through Knowledge Transfer 2012*, SIST 18, pp. 105–106.
DOI: 10.1007/978-3-642-34219-6_12

[3] Hein, L.: Impact of Research in Engineering Design on Industrial Innovation. In: International Conference on Engineering Design, ICED 2005, Melbourne, Australia (August 2005)
[4] The WIPO University Initiative: Developing Frameworks to Facilitate University-Industry Technology Transfer (December 2005)

Knowledge Transfer through Diversity Coach and Mentoring Partnerships

Ileana Hamburg

Institut Arbeit und Technik, FH Gelsenkirchen
hamburg@iat.eu

Abstract. Organisations recognise today that it is often difficult to recruit and retain personal with the necessary knowledge and to solve efficiently skill shortage i.e. by employment people with special needs. Mentoring and coaching particularly on the job under consideration of the diversity can be organised to address these aspects.

This article focuses on different aspects of knowledge transfer (KT), coaching and monitoring. The paper gives first a very brief introduction on diversity coaching (DC) and on formal and informal mentoring involving transfer of knowledge from more-to-less-experienced individuals; competences of a mentor in context of personal and professional requirements are also presented. A web-based community service approach by using social media is outlined, which supports KT through mentoring in Communities of Practice (CoPs). They are networks of individuals who share a domain of interest and knowledge about which they communicate (online in the case of Virtual Communities of Practice – VCoPs). In the last parts of the paper examples from two projects are given. The project IBB 2 (http://www.lebenshilfe-guv.at/unsere_dienste/eu_projekte/ibb_2_integrative_behindertenbetreuung) is a European Leonardo innovation transfer project aimed to support people with disabilities to entry and to be successfully integrated into professional life through DC and the mentoring approach. One of the aims of the project Net Knowing 2.0 (http://www.netknowing.com/) is to help SMEs to turn their daily work into a source of corporate learning for all their employees and to support KT by efficient using of informal learning and introducing a mentoring approach.

1 Introduction

Mentoring and coaching are human resources development processes often used to induct, introduce and guide staff into places of employment. Mentoring is considered as an old form of knowledge transfer [5], [20] [23].

Interest in mentoring has varied over time and has been affected by economic and social factors.

Now, also in connection to the economic recession, complex client requirements, rapid development of new technologies, skill shortage in some sectors, more knowledge is necessary as well as flexibility of companies referring to a fast familiarisation with new working environments and technologies.

R.J. Howlett et al. (Eds.): *Innovation through Knowledge Transfer 2012*, SIST 18, pp. 107–119.
DOI: 10.1007/978-3-642-34219-6_13

> "The only thing that gives competitive advantages to an organization is what it knows, how it uses that knowledge and how fast it can learn something new" [1]".

That means organizations should develop a strategy of learning, knowledge transfer (KT) [12] and use of it integrated in their work and business environments giving support "just in time" and competitive advantages.

One problem in this context is that if the organisations invest more in KT they risk increasing the job mobility of their employees, as they gain more competencies. So, it is necessary that organisations develop approaches which promote the KT [3], [17], [18] and also support retention of the staff [22] as well as integration of people with special needs into work. But they recognise that it is often difficult to recruit and retain personal with the necessary knowledge to solve also skill shortage problems and to employ people with special needs.

Mentoring and coaching on the job under consideration of diversity can be organised to address these aspects [21]. Many organisations establish mentorships for staff then new ones are hired or as a part of leadership development. Often individuals seek mentors to help them in their career.

Some authors use the terms mentoring and coaching as a single concept. In this paper we regard them as different forms to support training particularly on-the-job and combined mentoring and coaching as a social KT approach.

Mentoring is commonly used to describe a KT and learning process in which an existing member of staff guides newcomers or less-experienced people in a task to develop professional skills, attitudes and competencies.

Coaching is rather aimed at giving guidance to individuals or groups on the development of specific skills that are needed to be applied in a specific job environment.

Mentoring is a complex process involving not just guidance and suggestion, but also the development of autonomous skills, judgments, personal and professional master ship, expertise, trust and the development of self-confidence over the time [5]. Mentoring can be established for a number of reasons [14], [15]. The nature of mentoring is "friendly", "collegially". Mentoring also has to operate within professional and ethical frameworks. If it is possible it should remain voluntary and subject to mutual agreement. Mentoring is not just about solving problems. However, problems often underline a decision to seek mentoring.

On the job (or workplace) mentoring [16], [20] is the main topic of this paper. This mentoring is a learning partnership between employees for sharing information, transfer of individual and institutional knowledge and insight to a particular occupation, profession, and organisation. It includes the accompanying career advancement and natural support for a selected employee (mentee) i.e. through an experienced colleague (mentor), in order to achieve some present goals of mutual benefit for the mentors, mentees and organisation.

This kind of mentoring as a combination of a guided KT and informal learning process is a powerful experience, but the problem is that it is often only accessible to a few numbers of employees and its benefits are limited only to those few who fulfil the conditions to participate. Later we will present also formal or structured mentoring which takes mentoring to a next level and expands its advantages and corporate value beyond the mentor-mentee relation.

The mentors consider the mentees' resources and transfer their own workplace related knowledge and experience to the mentees in order to support them continuously in their professional environment and development. This mutual harmonisation also includes aspects of thought, social integration in the organisation and optimal professional resources utilisation of the mentees and their workplace fellows.

This article focuses on different aspects of knowledge transfer, of coaching and mentoring. First, a very brief introduction on DC and on formal and informal mentoring involving transfer of knowledge [4] from a more-to-less-experienced individual is given; competences of a mentor in light of personal and professional requirements are presented. A web-based community service approach by using social media is outlined which supports KT through mentoring in Communities of Practice (CoPs) [10], [25]. They are networks of individuals who share a domain of interest and knowledge about which they communicate (online in the case of Virtual Communities of Practice – VCoPs [2], [14]).

In the last part of the paper examples from two projects are given. The project IBB 2 (http://www.lebenshilfe-guv.at/unsere_dienste/eu_projekte/ibb_2_integrative_behindertenbetreuung) is a European Leonardo innovation transfer project aimed to support people with disabilities to entry and to be successfully integrated into professional life through DC and the mentoring approach. One of the aims of the project Net Knowing 2.0 (http://www.netknowing.com/) is to help SMEs to turn their daily work into a source of corporate learning for all their employees and to support KT by efficient use of informal learning and introducing a mentoring approach.

2 Diversity Coach and Mentoring

The diversity concept originally was developed in America in the context of the civil right movement emerging from Martin Luther King, but was soon broadly adopted by all kinds of bodies, initiatives and enterprise. Today no major company or other initiative should deal without solid Diversity Management in the employment sector.

Utilisation of diversity in the context of mentoring context – as a resource – greatly depends on well-interacting coaches for this topic (called Diversity Coaches, DC). The DC [27] has to identify the variety of relevant external and internal features, processes and upcoming developments influencing the performance of the enterprise and its staff and overall satisfaction. Moreover, in this context 'diversity' does not only refer to disabled people. Diversity is meant to include all kind of personal traits, characteristics, looks, etc. that can be distinctive between people. That means the DC could be responsible for people within the company that have a certain age (very young or very old), foreign people how are about to integrate into society or people with disabilities.

Within the context of our on-going projects, DC should train mentors and have to define an appropriate functional matrix to enable mentors in their dealings with citizens and individuals or groups of people in special need for support. This functional shaping demand must include various occupations or other outer demands in understandable, acceptable and meaningful terms. It must be in balance with the outer and inner individual settings of the mentor, the mentee, the colleagues at work, the employer and the organisation as a whole.

Mentoring has been used in Europe for a long time. Mentor was the man Odysseus entrusted his kingdom to when he went to the Trojan wars. In classical Greece, young men often lived with more experienced elders to learn not simply knowledge but, in addition, skills and attitudes. Mentoring relationship was also evident in the guilds of medieval Europe and the forms of apprenticeship that evolved from them.

Little research has explicitly explained mentoring as a means by which knowledge is transferred among individuals [6]. Some researchers started to examine the connections between mentoring and knowledge transfer.

In the IBB2 project we use the approach of mentoring on the job where the mentors, which are trained by DC, are companies' employees (see part 3 of the paper).

The advantages are obvious. The companies' employees know the work processes, which knowledge is needed for their efficiency and which the companies' knowledge resources are. They are used to the working environment and can estimate the hazards and situations which could be challenges to mentees or their fellows at the workplace. A further advantage is the development of a situation of mutual trust between the colleagues, which later is the basis in the daily work. In that way, the mentor acts in a twofold role being colleague on the one hand, and being mentor, on the other hand. Such a procedure ensures a smooth transition of the person from mentee to employee and an effective transfer of tacit knowledge acquired primarily through experience, not being easily expressed in words and numbers.

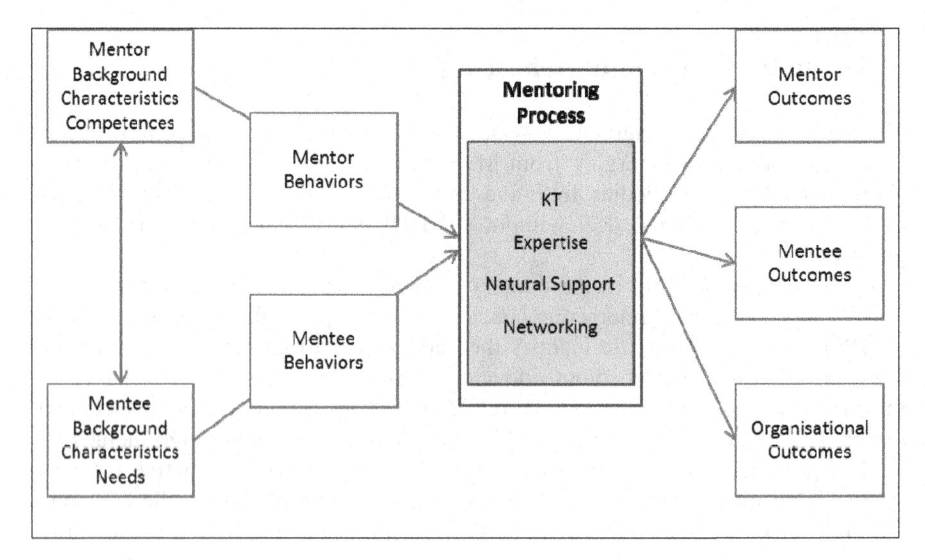

Fig. 1 An efficient mentoring process, Source: Giancola [7], IAT

The mentor on the job acts in the companies by supporting mentee with answering specific workplace-related questions or helping with the accomplishment of practical abandonments.

The following figure shows an example of an efficient mentoring partnership including also KT which we use in our projects.

Experience and expertise is necessary in the mentoring as well as being moderately extraverted. In the relation with the mentee, motivation and responsively is required. The mentee needs to be ready for professional development, open to learn and accept feedback. Time and initiative are necessary too.

The company can have benefits from mentoring by a quick introduction of the mentee into formal and informal company structures and demands, facilitating a deliberate, systematic and smooth transfer of technical or internal knowledge, opportunities to shape the workforce of the future in an international, deliberate way to meet company strategic goals and objectives, training of social competence of the mentee and the mentor.

Mentees have the opportunity:

- to meet with a trusted person to enter into a work place quickly and to cope with initial problems to discuss and resolve emerging job problems of genuine nature and in relation to the individual needs,
- to learn setting realistic goals and achieving them,
- to acquire new skills and enhance their skills and thus their future career opportunities and prospects in the future through the KT from the mentor,
- to build relationships or interactions allowing them to secure, maintain and advance in the job choosing a way that corresponds to the work routines and social actions of other employees,
- to receive (and contribute to) natural support, meaning: booming linked to existing social supports in the work environment.

For the mentors the training means the enhancement of their leading and counselling skills, development of their role within the company, possibility to share their professional experiences with others, exchange with other mentors by using, e.g., a web-based system [19] to support mentoring (WBMCS see part 3 of this paper).

In connection to the workplace fellows (team), the mentor has to inform and actively involve them with regard to the mentee and the various aspects of the mentoring process and its goals prior to the arrival of a mentee and constantly thereafter. Apart from knowledge transfer, communication with the team must include building solidarity, empathy and preparedness to actively take over responsibilities.

Additional duties of the mentor include: informing and actively involving together with the job coach the employer on the outcome of the mentoring process and the performance of the mentees. From the very beginning onwards, it is crucial to actively engage the employer into the mission and the various aspects of the mentoring process and its goals. The role of a mentor should include also building a positive outcome primed relationship between all involved, setting objectives/action plans, supporting organisation innovative changes and retention of qualified staff, recognising success, empowering/encouraging/motivating people with special needs, formulating expectations for all involved.

In the following, we present the two types of mentoring, informal and formal ones:

Informal mentoring aspects:

- Goals of the relationship are not completely specified
- Outcomes cannot be measured in total
- The process of KT cannot be explicitly described and it is based on the ability and willing for this process
- Access is limited and can be exclusive
- Mentors and mentees are often selected on the basis of personal chemistry, which means an initial connection or attraction between them
- Mentoring lasts a long time
- The organisation benefits indirectly, as the focus is exclusively on the mentee.

Formal mentoring aspects:

- Goals are established from the beginning by the organisation, mentors and mentees
- Outcomes are measured
- Knowledge which has to be transferred is known at the beginning
- Access is open to all who meet the criteria established by the organisation for the corresponding mentoring program
- Mentors and mentees are paired based on compatibility
- Organisation and employees can benefit directly.

Mentors need to be strongly focused, principled and able to develop empathy with the perspectives of others. To be a mentor especially requires having:

A. Social Competences (particularly interest, motivation, awareness, (verbal and non-verbal) communication, aptitude, empathy and engagement skills).
B. Professional Competences (responsibility standards in knowledge and skill).
C. Operating Competences according to ethical and professional standards and to know the boundaries when engaging with mentees.

In Europe and elsewhere, due to internationalisation, intercultural knowledge and competence in social and employment settings become more and more essential [8]. Intercultural competence does not refer to a set of methods and techniques with which one can change the own actions as culturally more appropriate [11]. It refers to a "holistic" aptitude and competence which is realised at all significant levels of the mentoring process and on top pays special attention to features like e.g. language, wordings and modes of expression, traditions, value of tolerance, legislations, religion, gender, nutrition individual time management and others.

The mentor-mentee relationship can be challenged by cultural diversity. Mentors can be trained to face such challenges successfully and with sufficient respect, tolerance and understanding, but avoidance of simple fraternisation and to transfer cultural knowledge to the mentees.

3 ICT Support of Mentoring and KT in a Community of Practice

We decided to support the training, mentoring and the KT process in two of our projects which we will describe in the next part of the apper through an ICT-based approach, a web-based one (WBMCS) allowing mentors, DC and mentees to learn on-line, to communicate and collaborate, and to transfer and share knowledge. Social media, particularly based on Web 2.0, i.e., media which supports social interactions and social KT, can be used to develop such systems. Social media can take many different forms, including internet forums, weblogs and wikis [26].

The technical skills needed to use social media are rather low. Blog software, for example, can replace sophisticated and costly content management systems. It enables content providers (reporters, writers, educators) to concentrate on their content without bothering too much about the underlying technicalities. It is even faster and less demanding to communicate through social networks, such as Facebook, Twitter and others. Another important feature of such applications and "spaces" is the decreasing differences, such as the one between teachers and taught, between formal and informal learning processes, between education and knowledge transfer. This gives rise to new integrate and world-wide forms of learning, e.g., in Communities of Practice (CoPs). Here, a community based on shared interests, learns in a community of equals (content wise, experience wise and truth assuring, however, non-equal participants) by exchanging expertise and transferring knowledge without building a hierarchy, because any of the participants is considered expert in a particular field, teacher and taught at the same time. A low-cost and easy access virtual room to accommodate formal and informal learning practices, group collaboration and the gathering and exchanging of learning materials might be realised in an e-Learning environment based on the social media tool TikiWiki CMS Groupware.

Web-based supported mentoring has a number of benefits including:

- Provision of a 24 hour access of saved knowledge, for training material and communication
- Accessible anywhere with internet availability
- Provision of a platform even if face-to-face communication is not possible
- Learning assessment and progress monitoring of the mentor-mentee relationship.

One important activity for the successful management of knowledge transfer within the mentoring process supported by a CoP is to define common goals for this process in advance. The goals have to be identified and agreed by all members. Other aspects are trust and the depth of relationships. Face-to-face interaction and socialisation processes consolidate the relations between members and group membership. Trust is important for knowledge sharing and development in a virtual team or virtual CoP and this develops primarily through face-to-face interactions. So we started the mentoring in our next described projects with face-to-face sessions.

An ICT supporting mentoring approach accounts for the varying learning abilities of students and overcomes the limitations in time or space etc. of traditional training environments which are restricted to rules in order to adequately fine tune a group to pre-defined criteria. The WBMCS, when adequately designed, can reduce the limitations of the classroom and allow the learner to work at his or her own space, speed and depth with structured support from both, the educators and the other learners.

However, for quality reasons and outcome value, traditional elements of monitoring/mentoring have to be affiliated. Experience from other projects demonstrates the need for a constant presence of experienced and qualified mentors in the WBMCS. A trust relationship has to be established online, if the WBMCS will be used in the project on a regular basis by the mentors and mentees. The WBMCS should support the motivation and retain students in the learning process. The WBMCS should be used in a context so that personal issues, which are not suitable for the online environment, do not become accessible for all. The WBMCS should support real mentoring and not be understood as a supervisory tool.

4 Examples

IBB 2 (Integrated Care Taking http://www.lebenshilfe-guv.at/unsere_dienste/eu_projekte/ibb_2_integrative_behindertenbetreuung) is an European Leonardo innovation transfer project aimed to support people with disabilities to entry and to be successfully integrated into professional life through DC and the mentoring approach. In this project, the main KT process to the DCs will be done by the project partners coming from five European countries and who trained the DC. DCs will transfer their knowledge to mentors during the training in all partner countries. So the task for the DC in the process of mentoring on the job of people with special needs (i.e. disabilities like in the IBB2 project - www.ibb2.com, senior requirements, migration background, etc.) is to transparently instruct and enable mentors acting for

- social inclusion and integration of people with special needs
- active participation of them according to their individual ability, character, temperament and talent development in line with general rights plus sustainable cultural need developments or mission statements (e.g. EUROPE 2020 visions) keeping the awareness of the employers' and employing institution's interests for contributing to social cohesion and respective values in society

The training of DC and mentors supported by a WBMCS prototype takes place by using a blended learning approach, i.e. starting with face-to-face training and using also e-Learning. For mentoring, we will use both methods: informal and formal ones. We use TikiWiki for our WBMCS prototype in the IBB2 project. This WBMCS prototype supports the following processes in a Community of Practice (www.ibb2.com):

- Training of the DC and mentors
- KT from DC to mentors and from mentors to mentees
- Mentoring process including also mentee learning.

The project team encourages the CoP members to use particularly the weblogs and forums supported by the portal which can contribute to transfer also tacit knowledge.

Fig. 2 IBB2 WBMCS
Source: www.ibb2.com

Training material is available online in the WBMCS, but in addition, a discussion forum for each module enables learners and trainers/experts to exchange and add ideas to the environment. This allows learners to provide feedback (anonymously, if desired) to the experts. It also enables them to pose queries to which other participants or the experts can answer. All participants are able to see the initial queries and the discussion stream of answers from other participants and the instructors.

Another EU on-going innovation transfer Leonardo project is Net Knowing 2.0: Web 2.0 Technologies and Net Collaborating Practices to support learning in European SMEs (www.netknowing.com). One of the aims of the project Net Knowing 2.0 is to help SMEs to turn their daily work into a source of corporate learning for all their employees and to support KT by introducing a mentoring approach. The integration of the e-Learning in the daily working life will be facilitated, as well as the opportunity for companies to build their own formative resources including knowledge basis. In the first year the project partners discussed KT elements that can be done from other projects. Then they discussed with SMEs about efficient using of informal learning also within a mentoring process. The goal of mentoring staff from SMEs is to improve job performance by increasing employee's capability to manage their own performance emphasing on trust, experience, and supervision, to facilitate performance and KT in the organisation, to support retention and leadership development. The mentoring approach is not used within SMEs in Germany, so a workshop has been organised to discuss with representatives of SMEs some tactics for implementing a mentoring program in their companies. One possible approach is that experienced at the Virtual Academy Brandenburg. The coaches and mentors are external persons who should support the sustainable development and advancement of strategic competences of SMEs through informal and organizational learning. Potential learning consultants can be trained to act as mentors. Two SMEs decided to try this concept and another three decided to use mentors which are staff in the company. In this case the KT will be very beneficial for the mentee in the own career but and also for the mentor. The mentor could have same benefits from the mentee and at the same time gains leadership skills by the act of mentoring. It is supposed that both individuals developed skills within this sponsored KT process and the company is providing a way for KT before an employee's retires or leaves the organisation.

At the workshop, SMEs from Germany proposed a route map for the successful deployment of a coaching and mentoring program within the specific context of a SME environment:

- Putting the specific working environment into context.
- Researching the role played by the organisational culture or "climate" in the development, maintenance and success of the SME.
- Determining real qualification needs of the staff before starting the mentoring process.

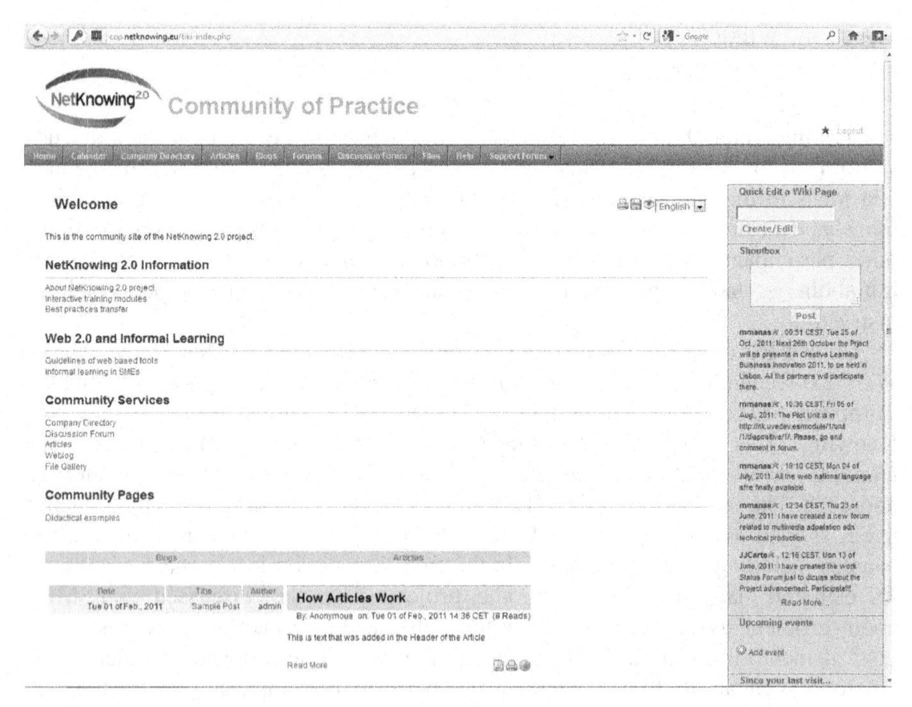

Fig. 3 Net Knowing 2.0 WBMCS
Source: http://cop.netknowing.eu

- Determining the knowledge gaps and which of them can be minimised by a mentoring system.
- Demonstrating that a mentoring intervention has real benefits in this context and not being bureaucratic.
- Being a process based on trust, experience, and supervision and informal learning.
- Identifying barriers to effective mentoring/coaching.
- Determining issues that need to be incorporated within the mentoring/coaching intervention, if it is to be successful.
- Qualifying coaches and mentors for different forms of working and learning.
- If necessary, using big companies for delivering mentors (learning consultants).

A first version of a WBMCS prototype supporting KT and mentoring within a CoP has been developed by using TikiWiki.

5 Conclusions

Mentoring, particularly in combination with diversity can be an excellent tool for retention, progress and integration of staff in a company/organisation if implement appropriately. It is a solution that can meet the needs of individuals of all ages and all levels in their work life and career. Additionally mentoring provides an efficient way for KT particularly supporting skill shortage and as the older generations start in their retirement. By supporting mentoring within informal and formal approaches, ICT methods should be blended with traditional face-to-face ones.

Acknowledgement. This paper describes work within the Leonardo innovation projects IBB2 and Net Knowing 2.0.

References

[1] Argote, L., Ingram, P.: Knowledge transfer: A basis for competitive advantage in firms. Organizational Behaviour and Human Decision Processes 82(1), 150–169 (2000)

[2] Diemers, D.: Virtual Knowledge Communities. Erfolgreicher Umgang mit Wissen im digitalen Zeitalter. Dissertation der Universität St. Gallen (2001)

[3] Duguid, P.: The Art of Knowing: Social and Tacit Dimen-sions of Knowledge and the Limits of the Community of Practice. The Information Society (Taylor & Francis Inc.), 109–118 (2005)

[4] Eby, L.T., Allen, T.D.: Moving toward interdisciplinary dialogue in mentoring scholarship: An introduction to the special issue. Journal of Vocational Behavior 72(2), 159–167 (2008)

[5] Edelkraut, F., Graf, N.: Der Mentor – Rolle, Erwartungen, Realität. Pabst Science Publishers (2011) ISBN 978-3-89967-723-2

[6] Gallupe, B.: Knowledge management systems: Surveying the landscape. International Journal of Management Reviews 3(1), 61–77 (2001)

[7] Giancola, J. (2011),
http://www.naed.org/uploadedFiles/NAED/NAED_Site_Home/Events/Market_Area_and_Niche_Meetings/Human_Resources_and_Training_Conference/3b%20-%20Giancola%20-%20Mentor%20Presentation.pdf

[8] Gupta, A.K., Govindarajan, V.: Knowledge flows within multinational corporations. Strategic Management Journal 21, 473–496 (2000)

[9] Hall, B.: Learning goes online: how companies can use net-works to turn change into a competitive advantage. Cisco Systems: Packet Magazine (2000)

[10] Hamburg, I., Engert, S., Petschenka, A., Marin, M.: Im-proving eLearning 2.0-based training strategies on SMEs through communities of practice. In: The International Association of Science and Technology for Development: The Seventh IASTED International Conference on Web-Based Education, Innsbruck, Austria, March 17-19, pp. 200–205 (2008)

[11] Hamburg, I.: Intercultural learning and collaboration aspects in communities of practice: poster. In: WEBIST: 7th International Conference on Web Information Systems and Technologies, Noordwijkerhout, The Netherlands, May 6-9, pp. 503–506. SciTePress – Science and Technology Publications, Lissabon (2011)

[12] Hamburg, I.: Supporting Cross-Border Knowledge Transfer through Virtual Teams, communities and ICT Tools. In: Howlett, R.J. (ed.) Innovation through Knowledge Transfer 2010. SIST, vol. 9, pp. 23–29. Springer, Heidelberg (2011)

[13] Hamburg, I., David, A., Deodato, E., Costigliola, V., Verbeek, A., Breipohl, W.: Social Services and interdisciplinary EHEA-ERA strategies for various target groups with special needs (IBB2). In: Eracon 2011/Athens 29 April/HASAC Workshop (2011)

[14] Johnson, C.: A survey of current research on online communities of practice. Internet and Higher Education 4, 45–60 (2001)

[15] Johnson, W., Ridley, C.: The Elements of Mentoring. Überarbeitete Ausgabe. Palgrave Macmillan, New York (2008) ISBN 978-0-230-61364-5

[16] Kram, K.: Mentoring at work. Developmental relationships in organizational life. Scott, Foresman and Company, Glenview (1985) ISBN 0-673-15617-6

[17] Krogh, G., Ichijo, K., Nonaka, I.: Enabling Knowledge Creation. How to Unlock the Mystery of Tacit Knowledge and Release the Power of Innovation. Oxford University Press, New York (2000)

[18] Nonaka, I., Konno, N.: The concept of 'ba': building a foundation for knowledge creation. California Management 40(3), 40–54 (1998)

[19] O'Reilly, T.: What is Web 2.0. Design patterns and Business models for the next generation of Software (2005), http://www.oreillynet.com/lp/a/6228

[20] Ragins, B., Kram, K. (Hrsg.): The Handbook of Mentoring at Work: Theory, Research and Practice. Sage Publications (2007) ISBN 978-1-4129-1669-1, LCCN 2007016878. Buchbeschreibung (eng-lisch) beim Verlag

[21] Richert, V.: Mentoring und lebenslanges Lernen. Individuelles Wissensmanagement im Informationszeitalter. Müller, Saarbrücken (2006) ISBN 3-86550-975-4

[22] Rousseau, D., Shperling, Z.: Ownership and the changing employment relationship: Why stylized notions of labor no longer generally apply—A reply to Zardkoohi and Paetzold. Academy of Management View 29(4), 562–569 (2004)

[23] Stephenson, K.: What knowledge tears apart, networks make whole. Internal Communication Focus, 36 (1998), http://www.netform.com/html/icf.pdf (retrieved December 3, 2008)

[24] Stocker, A., Tochtermann, K.: Investigating Weblogs in Small and Medium Enterprises: An exploratory Case Study. In: Proceedings of 11th International Conference on Business Information Systems – BIS 2008 (2nd Workshop on Social Aspects of the Web), Innsbruck (2008)

[25] Wenger, E., McDermott, R., Sydner, W.: Cultivating communities of practice: a guide to managing knowledge. Harvard Business School Press, Boston (2002)

[26] Wikipedia, `http://en.wikipedia.org/wiki/Tikiwiki` (retrieved January 03, 2010)

[27] Wingels, R.: Diversity Mentoring. Unterschiede erkennen, die einen Unterschied machen. In: Koall, I. (Hrsg.) Diversity Outlooks. Managing Diversity Zwischen Ethik, Profit und Anti-Diskriminierung, pp. 978–973. LIT, Münster (2007) ISBN 978-3-8258-9745-1

The Third Way for the Third Sector: Using Design to Transfer Knowledge and Improve Service in a Voluntary Community Sector Organisation

Mark Bailey and Laura Warwick

School of Design, Northumbria University, Newcastle-upon-Tyne, NE1 8ST, United Kingdom
{mark.bailey,laura.warwick}@northumbria.ac.uk

Abstract. This paper describes a two-year Knowledge Transfer Partnership that concluded in September 2011. Knowledge Transfer Partnerships (KTP) is a UK-wide activity that helps organisations to improve their competitiveness and productivity by making better use of knowledge, technology and skills within universities, colleges and research organisations. This paper details the outcome of a KTP between Age UK Newcastle and Northumbria University's School of Design that aimed to use Design approaches to improve the charity's services. This paper will describe the recent context for organisations operating in the Voluntary Community Sector and discuss the relevance of a Design approach to both the improvement of customer services in this circumstance, as well as the transfer of knowledge to a capacity-starved organisation. It will also document how Design was used to achieve both of these aims, and the resulting impact of this engagement on the organisation and stakeholders.

1 Introduction

In financially uncertain times such as these, navigating a clear road to service delivery is increasingly challenging. This is especially the case within third-sector organisations where funding, largely derived from central and local government and voluntary donations, is under considerable threat. However, with increasing pressure on voluntary organisations to help deliver vital services, it has become crucial that charities find sustainable solutions to deliver high quality customer service.

Age UK Newcastle, a charity that enhances the status and wellbeing of older people in Newcastle-upon-Tyne, has recognised the key role that a Design approach can play in helping them to creatively respond and pro-actively to this demand. Rather than engage an outside consultancy to help them to re-evaluate their service offerings, Age UK Newcastle appreciated the long-term benefits of embedding this knowledge within their organisation could bring.

Through the UK Government's Knowledge Transfer Partnership (KTP) scheme, Age UK Newcastle worked with Northumbria University's School of

R.J. Howlett et al. (Eds.): *Innovation through Knowledge Transfer 2012*, SIST 18, pp. 121–133.
DOI: 10.1007/978-3-642-34219-6_14

Design until September 2011 to transfer Design capability to the organisation. By embedding this knowledge within the very culture of the charity, Age UK Newcastle could improve their existing services, and also gain a skill base that would help them to develop high quality offerings in the future to continue to meet the needs of an ageing population.

2 Context

Whilst the recession has effected all corners of our society, the Voluntary Community Sector have been particularly impacted by the volatile fiscal climate. Over the past twenty years there has been a significant shift in the Voluntary Community Sector (VCS) landscape, 'from grant aid supporting charities… to them being contracted to do that work on behalf of statutory organisations' [1]. As a result of this increased reliance on public funding, the sector has found itself in an increasingly precarious state with the threat of significant reductions in public sector spending, as well as funding available from trusts and foundations.

In 2008, Age UK Newcastle, who were then known as Age Concern Newcastle, were one of the organisations that gained much of it's funding from local government. This is a common picture in the North East of England, where the VCS community has a disproportionate reliance on public money [2]. At this point, the senior management and trustees recognised the potentially hazardous position they were in, and began constructing an agenda to take them toward a more sustainable future.

Although this change agenda was already in full swing when the findings of the Spending Review (2010) were announced, the organisation still felt the full force of its consequences. The local government cutbacks that followed resulted in 73% of the VCS community in the region suffering a reduction in funding, this included Age UK Newcastle. More widely, the consequences of these actions have led to 40% of the region's VCS organisations making redundancies, and over a quarter decreasing the number of services that they provide [3], thereby having a considerable impact on the beneficiaries and communities they serve.

Despite this considerable reduction in capacity, the third sector community has been trying to cope with a sizeable increase in service demand [3]. However, the consequences of these cuts and an uncertain financial future is hampering VCS service deliverers' abilities to develop and deliver quality services. In such a dynamic operating climate, VCS organisations have been finding themselves without the time, resources or process to consider the underlying issues driving any service demand, and therefore the best way for them to respond to that need.

Given these recent pressures, there was an appetite within Age UK Newcastle to consider carefully the value of their current services and potentially reconfigure the way in which those services were offered. But in order to do things differently, they recognised the need to first perceive things differently, and after a chance meeting with a Design for Industry student at Northumbria University, they began to see that Design might provide a different lens through which to view well-established challenges.

3 Initial Project

In response to a Royal Society for the encouragement of Arts, Manufactures and Commerce (RSA) competition brief focused on reducing social isolation and loneliness in older people, a Design student had chosen to conduct a design project with the staff, stakeholders and customers of Age Concern Newcastle. Through this collaboration the student conducted first-hand research, accessing older people who at one point were lonely and/or socially isolated. From the insights gained they generated, developed and refined concepts that sensitively and appropriately addressed the needs identified through the research They also paid particular attention to the communication at all touchpoints[1] of the service to improve the customer's journey, as well as addressing the overall communication package to increase awareness and create a more accurate and positive picture of the services provided by Age Concern Newcastle.

In using a participatory approach, rather than an extractive one, the student created a service that was both appropriate, and of genuine value. The organisation had previously tried to tackle exactly this issue with traditional Marketing approaches without success, but through a Design approach, the service created was pertinent, sensitive and completely feasible. Age Concern Newcastle saw the potential for an iterative, participatory Design process to provide a key foundation for the organisation's ongoing change process, ensuring the customers' needs and demands were accurately represented throughout.

4 Design for Improving Services

Despite the term 'design' making more frequent appearances in policy and strategy documents within the Third Sector in recent years, the Design discipline is rarely formerly engaged to help develop services. Similarly, the need to engage service users in service development is a widely accepted necessity in the work of voluntary and public sector organisations, but whilst the concept may be embraced by many organisations, the practice is often not undertaken properly [4]. Outside of the Design community, there is often insufficient understanding of the role Design can play. Frequently thought of as a styling activity, non-designers fail to comprehend that the approach can help organisations to understand the needs and demands of their customers, and translate these into tangible outcomes.

Design has traditionally been viewed as a commercial activity. However, as the discipline has expanded, it has moved into a new arena of tackling social issues within a service context; creating systems and services to help people and society, as opposed to selling them for commercial gains [5]. The successes of programmes that use Design to tackle social challenges, such as Dott 07 [6] and

[1] Touchpoints are the elements of a Service as experienced by a real service-user and include a wide variety of components. Depending on the nature of the Service being developed, Touchpoints can include: waiting rooms, call centres, texts, bus journeys, pass cards, social networks, tickets, rewards, podcasts, referrals, feedback forms, payment receipts, rewards, queues etc as well as traditional marketing collateral.

Public Services by Design [7] have demonstrated on an international level that Design Thinking can make a valuable contribution to help tackle today's social and economic challenges [8].

Recent studies by scholars, practitioners and government bodies have suggested that Design has 'the power to stimulate or drive innovation and transform organisations and even societies' [9]. In the context of social issues, viewing an issue from a Design perspective has been said to bridge the gap between deductive and inductive thinking; using abductive reasoning to consider what could be [10]. In actively looking for new opportunities, challenging accepted explanations and inferring new possibilities, thinking as a Designer can help to visualise new ways of addressing well-established problems [10].

Adopting this creative perspective has been termed Design Thinking [11]; quite literally thinking as a designer would [10]. This very broad definition has been adeptly described by Tim Brown of IDEO, who states that it is a 'a discipline that uses the designer's sensibility and methods to match people's needs with what is technologically feasible and what a viable business strategy can convert to customer value and market opportunity' [11].

Design Thinking and it's abductive reasoning has been termed the 'third way' [12]; an alternative to a purely subjective, intuitive way of thinking, and a rational, analytical approach. In recent years, there has been much documentation of how this 'third way' can help organisations to respond to the challenges faced in times of austerity 'by thinking, and doing, differently' [13].

Design Thinking's disruptive, creative approach can help organisations to think in a radically different way, as opposed to taking small, incremental steps based solely on what exists. This made it an ideal approach to adopt at Age UK Newcastle when trying to address social delivery challenges in a more sustainable way.

5 Design for Knowledge Transfer

An important part of the Design Thinking approach that aligns itself with the VCS ethos, is its human-centered, participatory nature. Corrigan sees one of the distinguishing aspects of the third sector as its ability to work with its customers and empower them; often turning them from service recipients into service providers [14]. The ethos of a VCS organisation is not only to help address an issue, but to help people affected by that issue to add value to their own lives by being active in solving that problem [14]. Similarly, contemporary Design practice has moved from designing for people, to designing *with* people, and actively draws on the expertise and opinions of all stakeholders in the design process [9, 11, 15].

More importantly, Design Thinking sees the potential in everyone to be a designer [16, 17, 18] and places the tools and skills from the Design profession into the hands of these stakeholders, to help to co-produce the best service [11]. Design is therefore an appropriate approach to adopt in order to build on the participation already advocated within the organisation, and extend that to involve and capture insights from all of the stakeholders.

Likewise, the participatory nature of Design makes it an ideal vehicle for knowledge transfer. By actively involving stakeholders through a project process, they gain an in-depth understanding of the construction and use of tools and methodologies at every stage, as well as learning vital information about the services they currently provide and what their customers are looking for. The Design approach helps them to experience the application of the knowledge, and intimately understand its value, thus helping to fully integrate into organisational practice.

6 Culture Change

In September 2009, Age UK Newcastle embarked on a two-year Knowledge Transfer Partnership programme with Northumbria University to begin to capitalize on the Design knowledge that they had been introduced to. As part of that partnership, a new Design Graduate with excellent design and interpersonal skills was carefully selected and placed into the organisation to orchestrate this change from within. The graduate was supported by an academic team from Northumbria University School of Design.

The first year of this project was planned to contribute towards the overall strategy of the organisation, and define actions to be taken over the coming periods to ensure that they were meeting their customers' needs. To contribute towards this, the designer devised a project structure designed to effect culture change.

In order to achieve this cultural change within the organisation, it was essential to engage all stakeholders as quickly as possible. The first challenge was to help all the people in the organisation to fully understand the value of Design Thinking and how it promotes better customer experience. This was hindered by the levels of understanding of the term 'design', magnified when coupled with applying this in a service context.

By introducing examples of conceptual service developments that had been created in the context of their work, it was possible to make Design Thinking tangible for stakeholders. Using the student project that had started the organisation's design engagement helped people understand that the process was (in this case) about simple, sensible innovation to improve what was offered, not drastically alter it. Familiar examples of Design from popular television documentaries also played a role in demonstrating the breadth of 'design' and the role that the non-designer can play. Examples such as the popular UK TV chef Jamie Oliver and the retail expert Mary Portas were used to show that design is not always about 'making' something, it can be about adopting a different approach.

Whilst Age UK Newcastle had a commitment to improving services, it was essential that strategic culture change happened simultaneously. Age UK Newcastle has traditionally been a top-down organisation. However, as part of the Design approach knowledge would be gleaned from front-line staff, and they would be actively involved in making strategic decisions. It was therefore important that the senior managers embraced this practice, and reflected it throughout the organisation.

To this end, the KTP team devised a programme for senior managers to meet on a regular basis to share current work and develop their design skills. Named 'Ideas for Change', it also became a forum for discussing organisational issues like the financial structure of the charity, to add additional layers of knowledge that helped people shape ideas and give context to the overall change process.

Similarly, staff meetings were reinstated, which had dwindled during the previous two years. This two-year hiatus had resulted in a lack of communication and cohesion in the team; departments worked in silos and were not connecting their activities together, impacting on the unity of Age UK Newcastle's services. By developing a regular staff meeting programme, there was a platform for all staff to discuss work or share ideas, and this provided a channel of communication for changes happening in the wider world, as well as those being instigated through the senior management series.

Both 'Ideas for Change' and the staff meeting programme helped to encourage cross-service communication and working on ideas together to develop a more cohesive suite of services. It also helped to shift staffs' focus from their individual task and department, to seeing themselves as part of a larger organisation, and where their role contributed on this macro scale.

7 The Importance of Process

Another outcome of this new staff programme was recognition of the speed at which services were currently being designed, developed and implemented. Many third-sector organisations are, necessarily, reactive to funding opportunities, but this can lead to short-term, unsustainable services being put in place (and then removed when the funding has run out) without due consideration to the overall impact on strategic direction, brand, or most importantly, service users.

To help mitigate against the negative effects of this reactive style and based on a development of the UK Design Council model [19], a Research, Test, Refine, Implement, Review approach to new service development was introduced. The team also underpinned the importance of using customer experience to inform which direction to take at every stage, which helped support the shift in emphasis towards becoming a service-focused organisation.

In instilling the importance of process into the organisation, staff at all levels were taught how to prototype ideas and to refine, where necessary, before launching in order to reduce risk; this can require managing enthusiasm and reviewing the business and user case. It is particularly important to test ideas in this sector, where there is financial uncertainty and a vulnerable clientele. The design approach reduces this risk by actively prototyping and testing ideas for service elements before selecting the most effective solutions for detailed development. The KTP team introduced the knowledge and expertise to develop feasible prototypes that were necessary to elicit reliable service-user feedback.

To support the stages of the process, a toolkit was developed which detailed the importance of each stage, as well as providing a series of methods and tools that could be used to help to produce the best possible outcome. Each example was clearly explained in familiar language, and included a description of where it worked well, and how it should be managed. This put the tools of the trade into the staffs' hands to empower them to use the approach as part of their everyday activity.

8 Connecting with the Customer

As part of this first year programme, the design team also focused on aligning organisaitonal policy with customer needs. Unusually for most Design challenges, Age UK Newcastle were very empathetic to the needs of their customer; they had established relationships with many of them that meant they could offer, to some extent, a tailored service. Whilst in most organisations, the problem can be helping the staff to associate with the emotions of their customer, conversely in smaller charity groups, they can know their clientele too well. Knowing the specific circumstances of individual service users had coloured the way in which staff responded to service development; "we can't do that, Ethyl couldn't come on a Tuesday". Staff therefore focused on their current customer base when developing services, and consulted with them primarily when evaluating service offerings, providing a distorted representation of their success.

To help staff to see the bigger picture without losing sight of the individual, staff were involved in the facilitation of workshops to experience how best to gather true opinions from both existing and potential customers. Staff were instructed on the development and use of personas. This provided them with the means to develop archetypal characters that were used to support service developments in general without becoming too focused on the detail of an individual circumstance [20, 21].

To further bridge the gap between the staff and older person's perspective, all staff were invited to take part in a observational exercise around customer service. Named a 'Staff Safari', it was designed to encourage members of staff to think as a customer, and explore the different experiences on offer. Each member of staff was given a disposable camera and asked to take photographs that represented 'good' and 'needs to be improved' customer service, both within, and external to, the organisation. The photographs acted as a bridge between what the staff demanded as customers, and what they should be aspiring to provide within their own working environment. Discussing the ideas also provided staff with a platform to express their views and opinions about all parts of the organization; an opportunity they had not previously been afforded.

A Design approach addresses the whole person, the whole system, the whole configuration of people and tasks; it helps individuals to see services in the round. Operational staff tend to focus on what they have control over, or what they think they have control over. Whilst the activity had helped staff to make suggestions about other areas of the organisation, it also helped them to consider the user journey that occurs before or after users engage in the core service activity. In using a camera to capture issues that affected customer experience, they began to recognise the importance of each stage of the customer's experience, including their journey to the building and their first contact with a member of staff. The KTP team used staffs' photographs to create a photographic customer journey, demonstrating visually where customers might encounter barriers to accessing a service. This provided staff with the opportunity to look beyond their immediate activities and see the service more holistically.

The 'Staff Safari' also helped to give staff permission to make changes, and demonstrated that small adjustments that take little time or money can have a big impact. Involving staff directly in this design activity demonstrated the value of the tool in gathering insights, and providing a vehicle to contribute ideas and opinions. It also started the whole organisation collectively on the path to making changes towards improving service experience.

9 Carving a Path

During the KTP project, Age UK Newcastle was continuously affected by changes occurring in their dynamic operating context. Although the management had carefully planned towards more sustainable practice, the organisation's activity was threatened. As the project entered its second year, the organisation was in a particularly difficult period, staff found it increasingly difficult to focus on innovation. However, the KTP team recognised the importance of continuing to follow the Design approach, as customer-focused improvements would help to ensure the long-term future of the organisation.

To this end, the KTP team decided to conduct a design project within one area of the organisation; the Befriending Service, which provides social contact for isolated, lonely older people. Currently working at capacity, and with an uncertain funding future, it was an excellent role model to review in full view of the organisation; providing something to imitate in order to affect real system change.

The Befriending team were very keen to review their service offer, as they had a significant waiting list that they wanted to address. However, they were very solution-focused and wanted to devise options for expansion as soon as possible; they interpreted the demand for the service as effectiveness. They also gave little consideration to the efficiency of the offer.

To re-balance their perspective, staff were asked to articulate, and attribute a time to, every activity undertaken when delivering this service. A simple task, but it highlighted that the administration of the service was absorbing the majority of the coordinator's time. A budget analysis also revealed that the service had an organisational cost of £900 per service user, per year, a figure far higher than the team's estimates. These initial steps created amongst the team a motivation to address the inefficiencies in the service laying the foundations for undertaking a thorough research process.

10 Thoroughly Designed Research

The Befriending department previously conducted limited, infrequent research, considering it to be too costly for one so resource-strapped. The research it did conduct asked closed questions about the service and any improvements users would like to see. This structured style of evaluation provided little feedback and of poor quality, providing no guidance regarding the service's actual performance.

To prove the value of appropriately designed research, the KTP team wanted the staff to conduct as much of the activity as possible. They trained them in design research methods, so that they could understand the purpose of the techniques, and ensure that they collected the information in a consistent and appropriate way. Each activity was also designed with the staff team, focusing on the needs of the participants, and how best to elicit the information whilst providing an enjoyable experience for them. In constructing the activities with the team, they were shown that research could be thoroughly designed, and yet remain flexible and work within the organisation's ethical guidlelines.

First, existing recipients of the Befriending service were invited to come to an event to share their experiences and opinions. To prompt discussion, two fictional characters were introduced and the participants asked to suggest things that could improve the characters' quality of life. By creating characters in familiar circumstances and asking participants to consider what they may need and how they could be helped, this allowed people to think about their own needs without feeling embarrassed in the group setting.

To elicit the opinions of people on the waiting list, and those who were unable to travel, participants were visited in their own homes and semi-structured interviews were conducted. Participants were asked to complete a diary sheet to share with the team what they usually did on a day-to-day basis, and this formed the basis of an interview; allowing the researchers to tease out the emotions they experienced without prying into their personal life. Interviewers were also given conversation tips, feeder questions and visual prompts to help them be truly responsive to the participant, whilst gathering the necessary information.

Throughout the research process, researchers were asked to gather images of the participants or things that were of value to them. The service team were initially reluctant to do this, citing the vulnerable nature of the participants as a reason not to capture any information. The KTP team encouraged them to make this departure from their usual practice and ask each participant individually; respecting their decision, rather than making one for them. In practice, what they found was that almost every older person agreed to be filmed and photographed, and actually enjoyed the attention being paid to them and their belongings, making for a rich, interesting dialogue.

The photographs were used to create profiles of each older person and capture their own personal story. This format helped to gather data regarding family and friends, their typical week, and hobbies and interests in a visually stimulating way. The profiles also helped the staff remember details of interviews, inspired them to create solutions for real people, and helped to communicate effectively the content of interviews with other team members.

11 Doing Things Differently

Third sector organisations often work in partnership, but different funding bodies and management guidelines mean there is rarely comparable research data.

It is therefore not common practice at Age UK Newcastle to gather and consider all data simultaneously, but this is exactly what the KTP team did!

The process helped the team to pinpoint commonalities and differences in their findings, threading the information together to form a more cohesive understanding of what they had discovered. In examining the information as a whole, staff were also able to draw some conclusions based on in-depth research, rather than generalisations. The findings were translated into four distinct areas that needed to be addressed: connecting people with genuine friends; customer progression; enhancing the existing offer and more volunteering roles. These were used to inform the idea generation stage.

Where there had been no research, there had also been no idea generation; Age UK Newcastle had often emulated good practice taken from elsewhere without validating whether it was appropriate for their aim, customer or circumstance. The previous stages had provided insight into the changing needs, aspirations and interests of older people, and the KTP team wanted to use the research findings to generate numerous potential service innovations to demonstrate that those needs could be addressed in different ways.

The team were asked to consider not how they could improve their service, but simply how they could fulfil its aim: offering social contact to isolated older people. In rewording the question, the KTP team gave the service team permission to think broadly, creatively and differently. Staff found that by focusing on the potential of an idea, and withholding judgement, they could take inspiration from each other and produce surprising yet appropriate suggestions. Reframing the question gave the project team permission to think broadly, creatively and differently. The profiles were then used to inspire this process, and ensure the KTP team created solutions for real people; drawing out the important information to help develop new ideas that would address their needs.

The generated ideas were then shared with staff members from across the organisation to get their opinions on which ones should be a priority to develop. From this feedback, and the knowledge gained through the previous stages, the team developed a 'Telephone Neighbourhood' concept. It suggested a way of connecting customers with other customers by forming a 'neighbourhood' that contact each other by telephone every week, whilst the group is supported by a volunteer who helps them to make connections and develop friendships. It was suggested that once the network was established, the volunteer would gradually withdraw and the network would then self-sustain, making it a much more manageable option for the organisation in the long-term.

The concept was well received by staff across the entire organisation, which was important to ensure that it was considered an organisational initiative, rather than being segregated as a departmental offer. Additionally, some of the other generated ideas also provided inspiration for other departments and inspired them to review their current offers to older people and create more appropriate options.

12 Testing the Water

Being such a small organisation, Age UK Newcastle had never placed much emphasis on prototyping ideas before launching them. This has often led to unforeseen issues that have proved costly to the charity. Having instilled the importance of process across the charity, the team understood the need to test the 'Telephone Neighbourhood' concept in order to judge whether or not it was an appropriate response to the research findings.

To this end, the service was piloted with a control group to check that it operated as intended. The monitoring and feedback process was carefully designed to show whether the service was both effective and efficient, and giving the team an opportunity to refine the model before launching it full scale. At the time of this paper, the pilot was still being undertaken, but with positive initial feedback.

13 Wider Implications

Whilst the service review had resulted in a service innovation, there were also wider implications to the research findings that the design team wanted to share.

The design team developed a blueprint model for an ideal social care service experience, based on the results of the research. It responded to the findings that customers were not being accurately referred into the organisation, and instead were being basically assessed for a service i.e. a lonely person who was referred to Befriending by Social Services would be assessed for that service, not for any others that they might also be interested in. The blueprint highlighted the need to create a more holistic plan for a customer that would help them to reach personal objectives i.e. confidence building, in order that they would become a service provider, and not just a service recipient.

The model was shared with a wide group of stakeholders in the sector to much acclaim. By linking the more generic model to real client and service issues, the work had a more profound impact on their thinking, appealing to them on an emotional as well as professional level.

The service review was incredibly successful on a multitude of levels, resulting not only in an improved service offering, but also organisation policy that reflects the needs of the client group. The project work also inspired the team to apply this Service Design approach across the rest of their department. As a result, Age UK Newcastle are re-assessing all of their services and possible development options, and also seeing a Service Design approach as crucial to those reviews.

14 Conclusions

This programme has had a wide-reaching impact that has made Age UK Newcastle more customer-focused, more sustainable and more responsive.

The Knowledge Transfer Partnership team have found that using a Design approach has enabled managers, staff and service recipients to engage in service development in a different way by going on the journey together. The process provided a safe space for constructive feedback, opportunities to understand the subtleties of expectations and perceptions, and an approach for testing out new ideas as part of the design and development of services. It has been shown that Design offers both a rigour and creativity to service development, and complements more routine forms of engagement such as surveys, audits or focus group discussions.

The use of images and imaginative presentation as part of the Design process has been very effective in enabling people to get quickly to the heart of the matter. In many existing engagement processes, the use of this type of imagery, be it photographs, video or illustration, may be regarded as a luxury rather than a necessity. However, the team have shown that visualising an idea, process or touchpoint has a profound impact on a stakeholder's ability to understand the content, and also their likelihood to contribute feedback.

Our experiences have shown us that effective organisational change can be achieved by having someone in-house driving the change, as opposed to an external consultant influencing it. As an employed member of staff, the designer was a constant resource to help support the next steps of the organisational change, engaging stakeholders at pertinent times during their day-to-day activity, gradually educating them in Design Thinking methodology. They were in an ideal position to learn about, predict and respond to the changing contexts to produce a truly responsive approach, which is key for Third Sector organisations.

At a time when VCS organisations, and therefore the people that they help, are particularly vulnerable, the need for evidence-based practice becomes more important. The authors feel that Design Thinking offers a rigorous approach that provides the evidence base for service re-design and development, as well as the tools with which to embed this imaginative way of working.

The legacy of this project has been a culture-shift where service experience is at the core of the organisations work and staff are empowered to explore and initiate new opportunities (in a methodical way). We feel this would not have been achieved without the focus on practically using this new knowledge whilst simultaneously embedding it.

References

1. Bruce, I.: Kicking charities while Serco profits isn't a plan with legs, The Guardian (August 3, 2011), `http://www.guardian.co.uk/commentisfree/2011/aug/03/cutting-charities-funding-serco` (accessed: November 15, 2011)
2. Northern Rock Foundation, Trends in the North: what we have learned from the quantitative programme of the Third Sector Trends Study (2010), `http://www.nr-foundation.org.uk/resources/third-sector-trends-study/` (accessed: October 12, 2011)
3. VONNE, Surviving or Thriving: Tracking the impact of spending cuts on the North East's third sector (2011), `http://www.vonne.org.uk/news/news_article.php?id=1582` (accessed: November 20, 2011)
4. Douglas, B., as quoted in Warwick, L.: Designing Better Services Together. School of Design, Northumbria University, Newcastle (2011)

5. Parker, S.: Social Animals: tomorrow's designers in today's world (2010), http://www.thersa.org/projects/design/reports/social-animals (accessed: December 12, 2010)
6. Thackara, J.: Wouldn't it be great if... (2007), http://www.designcouncil.org.uk/publications/Dott07-Manual/ (accessed: November 3, 2011)
7. Design Council, Public services: revolution or evolution? (2010), http://www.designcouncil.org.uk/our-work/insight/public-services-revolution-or-evolution/ (accessed: November 3, 2011)
8. Schaeper, et al.: Designing from within: Embedding Service Design into the UK's health system. Touchpoint 1(2), 22–31 (2009)
9. Blyth, S., Kimbell, L.: Design Thinking and the Big Society: From solving personal troubles to designing social problems. Actant and Taylor Haig, London (2011)
10. Martin, R.: The Design of Business: Why Design Thinking is the Next Competitive Advantage. Harvard Business Press, Cambridge (2009)
11. Brown, T.: Change by Design: How Design Thinking Transforms Organizations and Inspires Innovation. HarperBusiness, New York (2009)
12. Brown, T., Wyatt, J.: Design Thinking and Social Innovation, Stanford Social Innovation Review (2010), http://www.ideo.com/images/uploads/thoughts/2010_SSIR_DesignThinking.pdf (accessed: November 15, 2011)
13. Runcie, E.: Can design be the answer to delivering quality public services in an environment of severe funding cuts? (2010), http://publicsectorinnovation.bis.gov.uk/can-design-be-the-answer-to-delivering-quality-public-services-in-an-environment-of-severe-funding-cuts (accessed: November 20, 2011)
14. Corrigan, P.: Saving for the NHS: The role of the third sector in developing significantly better health care outputs for the same level of resource. ACEVO, London (2010)
15. Sanders, E., Stappers, P.: Co-creation and the new landscapes of design. CoDesign 4(1), 5–18 (2008)
16. Simon, H.A.: The sciences of the artificial, 3rd edn. MIT Press, Cambridge (1969)
17. Buchanan, R.: Wicked problems in design thinking. Design Issues 8(2), 5–21 (1992)
18. Thackara, J.: In the bubble: designing in a complex world. MIT Press, Cambridge (2009)
19. Design Council, The Double Diamond (2005), http://www.designcouncil.org.uk/designprocess (accessed: November 3, 2011)
20. Blomquist, A., Arvola, M.: Personas in action: ethnography in an interaction design team. Paper Presented at NordiCHI 2002, Arhus, Denmark (October 2002)
21. Pruitt, J., Adlin, T.: The persona lifecycle: keeping people in mind throughout product design. Morgan Kauffman, San Jose (2006)

Living Labs: Frameworks and Engagement

Maurice Mulvenna and Suzanne Martin

TRAIL Living Lab, University of Ulster, Northern Ireland, UK

Abstract. This paper introduces the concept of living labs and shows the results of a survey of the living labs network. The main value of the study is that it provides findings about the diversity of living labs, how they engage with users and how strong the relationships are between living labs.

1 Introduction

The architect and academic, William J. Mitchell, created the concept of living labs. Mitchell, based at MIT, was interested in how city dwellers could be involved more actively in urban planning and city design (Mitchell, 2003). The ideas of citizen involvement in the design process was subsequently taken up and developed further in Europe by various research communities. A small number of living labs, created across Europe in 2005, primarily from the Computer Supported Cooperative Working (CSCW) research community, formed the European Network of Living Labs (ENOLL) in 2006. Successive waves of new living labs have since been created and, in 2011, there are, for example, 15 living labs in the UK and over 250 living labs across Europe and beyond.

This paper describes the results of a major survey of living labs, undertaken in 2011 (Mulvenna et al. 2011). This is the first major survey of the living labs themselves and the findings outline how living labs perceive themselves in terms of, for example, engagement with users, focus of work, future needs and financial position.

2 Living Labs

The ENOLL living labs recognise, as did Mitchell, that technology, in particular ICT plays a powerful catalytic role in user engagement and most of them are focused on using technology to support user engagement, research novel ways of engaging with users, and communicate findings rapidly and accurately using low-cost, mass-adopted tools such as social networks.

It is apparent from an examination of the living labs that many have a particular niche in which they operate. Some labs are region-based, others focus on a particular product family for example, automotive design, while others seek to address particular societal needs in, for example, healthcare. However, the use of technology to engage and support users as early as possible in product and service development is the common denominator for all of them.

R.J. Howlett et al. (Eds.): *Innovation through Knowledge Transfer 2012*, SIST 18, pp. 135–143.
DOI: 10.1007/978-3-642-34219-6_15

3 Survey Method

The survey was designed to establish basic information about the living lab phenomenon, which was 'born in the USA' but developed in Europe and beyond under the aegis of the European Network of Living Labs (ENoLL). As of October 2011, when the survey was completed, there were 274 living labs in existence. The survey was launched on 20 June 2011 and sent via email message to the contact details of 195 living labs, drawn from those extant 212 living labs from the first four waves where contact details could be ascertained and verified. The survey response rate was 28.7%.

4 Survey Results

The survey explored different topics including domains of activity of living labs and aspects relating to territory of the labs, before examining the users of the labs. The models used by living labs to engage with users were questioned, relating to techniques for engagement. The translation of results from engagement with users was the next focus of the survey, before how living labs are evaluated was considered. The 'user experience' in living labs was then surveyed before respondents were asked about the area of stakeholders in living labs and how living labs could be designed to be sustainable in the longer term. Finally, questions relating to how living labs interact with other living labs were asked before final questions exploring the financial models used for sustainability was surveyed.

4.1 Domains of Activity

The initial question asked the labs to say which area or domain best describes the activity of their living lab. The responses were based upon the classification used by ENoLL, encompassing Digital Cities, E-Manufacturing, Energy Efficiency (aka Smart Energy Systems), E-Participation, Future Media and Content Delivery, Health and Wellbeing, and Tourism. It was interesting to note that 'other' was the response with the highest value. This may indicate that the classifications used in ENoLL are not representative of the domains in which the living labs practice, apart from the domain of 'Health and Wellbeing' selected by over a quarter of respondents. However, respondents had to choose a single domain to describe their activity and perhaps those living labs that operate across several domains selected 'Other' instead of picking the most representative domain in which they work.

4.2 Territory

The architect and academic, William J. Mitchell, created the concept of living labs. Mitchell, based at MIT, was interested in how city dwellers could be involved more actively in urban planning and city design. The ideas of citizen involvement in the design process was subsequently taken up and developed further in Europe by various research communities, primarily ENOLL.

When asked if the activity of their living lab was specific to their region, their country or was international in scope, a clear majority selected 'Regional' (58.9%), while 33.9% selected 'International' and only 7.1% chose 'National'.

The responses to this question revealed what was believed only anecdotally beforehand, which was that living labs primarily operate at a regional level. This may be related to their genesis at a regional level, often within academic and research organisations, which will be examined in subsequent sections in this report.

Only a relatively small number selected 'National', indicating perhaps the minimal role in the development of living labs by national governments in Europe and beyond.

The European Commission provide implicit support to living labs by, for example, facilitating many living lab activities at practical as well as policy levels. The Commission also provide tangible explicit support, primarily in, for example, the incorporation of living lab methods and techniques in RDI calls for funding. This support by the Commission may be the reason for just over a third of the living labs indicating that they operate at 'International' level (33.9%), where international perhaps translates as 'transnational activities' between European organisations who have already formed partnerships through RDI funded activities.

4.3 Membership and Status

The living labs were asked about the legal status of their lab. It was anticipated that academic and research organisations may host many living labs and that this would be reflected in the answer to this question. While 28.6% gave 'University' as a response, 30.4% of respondents indicated that the public sector hosted their living lab, breaking down as 'Government' (10.7%) and 'Other Public Sector Organisation' (19.6%). The unexpected response was that 16.1% of living labs have a legal status as 'Private Sector Organisations'.

The majority of those who gave 'Other' (25%) as an answer for the legal status of their lab were labs formed as public-private-academic partnerships, under the triple-helix model outlined in the introductory section of this article.

4.4 Users - Their Engagement and Involvement

When asked approximately how many end users were involved in their living lab, there was a broad range of responses. Twenty-three living labs indicated that they involved 1-100 users (41.1%), while fifteen indicated the involvement of 101-1,000 users (26.8%) and eighteen indicated over 1,000 users involved in their labs (32.1%). One respondent indicated that they involved 1.4 million users, which may be feasible in this age of research using social media, but such figures should be treated with caution.

The next question put to the survey respondents asked how easy or difficult it had been to engage with end users, and 55.3% answered that it was easy or very easy while 44.7% answered that it was difficult or very difficult.

4.5 Indicators for Impact

In order to gain some insight about the indicators used by the living labs, the respondents were asked which indicators were used to measure the impact of their living lab. Many respondents (23%) said that they were not currently using any indicators, primarily because their lab had not yet set these up.

Several labs responses included business like comments such as: "Profit", "commercial success, number of ideas, industry involvement", "returning customer, confirmed subscriptions to our partnership, revenue increase to customers, feedback collected through word of mouth", and in one case it was clear that the indicators of project funding organisations were being used "Will be subject to ERDF rules - jobs/SMEs supported" and "number and amount of external funding". Several responses related to intellectual property indicators, including: "Number of spin-outs, patents, products"; however, it was revealing that such indicators were cited by a small number (less than 4%) of respondents.

Many labs cited more academic measures of performance including: "number of master theses, number of papers…", and a large number also cited general measures from academia, business and society, for example: "research output of the postgraduate researchers involved; - buy in and interest level of the community to the initiative; - interest of surrounding communities, government, and industry in our work".

However, the most common indicator cited by the respondents related to the engagement with end users: "Number of cases, number of end users, number of Living Lab projects", "Number of tests performed, Number of external customers, Number of end-users engaged", "The number of projects, and the number of users involved", "user satisfaction",

4.6 Translating Results

In order to understand more about the processes involved in engaging with users, the respondents were asked how easy or difficult they found it to translate results or feedback from end users into actual service or product change.

A clear majority of living labs (60.7%) answered that the translation process from end users to products or service change was difficult or very difficult, while 39.3% said that it was easy or very easy. While the response was expected in that a majority found it difficult or very difficult; it could be considered surprising that as high a percentage as 40% said that it was easy or very easy to carry out this fundamental process.

It was interesting to note that some respondents found that the reason was a kind of 'lost in translation' effect, for example: "it can be hard to get constructive and instructive comments from users that are not used to giving feedback and analysing a service working with small numbers in focus groups gives rich data but what if those statements are not reflected by the survey majority technical developers have own ideas and feasibility of request might not be given within a project timeframe and budget", "End user and engineers are not talking the same 'language' i.e., it is not always easy to understand what end user means and vice versa", "there is a need of translation, the language used is different when discussing with an expert or a layman", "Translation from end user to specialists" and "End users and developers are not speaking the 'same language'".

It is interesting to note that some living labs tackled the 'lost in translation' effect mentioned above by tackling it head-on, for example: "By involving the developers in the user-activities it is more easy to transfer the user feedback", "If the feedback is captured in a structured manner then there is usually a clear way how to translate it into service/product improvement", "Because the users and the designers are the same group", "...because we gather users' feedback in a way that allows us to modify the service easily taking in mind this opinion. We try to guide the feedback sessions in a practical way", "We are currently doing research on this translation phase. When the analysis of user feedback is done collaboratively, it is not that difficult" and "Once they get to involve in the project it's more easy to get the results and the feedback".

4.7 Evaluating Living Labs

The survey then asked several questions examining if living labs found it useful to have access to practical advice and assistance in several areas. On average, around three-quarters of respondents answered that they would find it useful to have access to advice and assistance in a variety of areas including 'Getting users interested', 'Getting end users involved in a practical way', 'Communicating the concept to end users', 'Getting end users to see the benefits' and 'Involving all end users rather than specific groups'.

Respondents were then asked about evaluation processes or procedures employed to learn how users view their experience of being involved in their living lab. In total, 73.2% of respondents answered that they did have some form of evaluation process in place, ranging from surveys to meetings or including both formal and informal processes, while 26.8% had no processes in place. This was perhaps a surprising result in that, more than other entities, living labs would be expected to ask their users about the experience given that the philosophy of living labs is all about engagement and evaluation.

For those respondents who did ask for feedback from users, the users' feedback was overwhelmingly positive or very positive (91.1%) with only 1.8% giving negative feedback.

The living labs were then asked if they would find it useful to have access to evaluation and research training to assist with evaluating user involvement. 58.9% answered that they would find it useful to access evaluation and research training to assist with evaluating user involvement and only 17.9% said that they didn't need access to such resources. So, while (from earlier) 73.2% of respondents answered that they did have some kind of formal or informal evaluation process in place, a majority still would like access to support for evaluation and research training to assist with evaluating user involvement. This indicates that living lab respondents perhaps would benefit from techniques such as peer benchmarking with other living labs or access to training resources in order to understand better the practical aspects of evaluation of users.

4.8 Stakeholders

The survey then asked if organisations including government, other public sector organisations (e.g. Local Council), universities, private sector organisations and the European Commission were involved in the delivery of their living lab. This question partly relates to the earlier question asking about the legal status of the living labs where around two-thirds answered that their living lab was governed by a public sector organisation of some kind, but the primary purpose for asking respondents this question was to learn about the degree of penetration of triple-helix partnerships in the stakeholder mix for living labs operational activities.

The responses indicate that universities (78.6%) and private sector organisations (75%) are well embedded in the activities of living labs. The high value for 'Other public sector organisations' (71.4%) and the low value for 'Government' (44.6%) reflects the answers given earlier in relation to the question on the territorial specificity of the living labs, where almost two-thirds of living labs answered that they operated at a regional level while only 7.1% operated at a national level-perhaps more evidence that living labs are a phenomenon that operate more at a regional level, often with local councils. The significant impact by the European Commission (24%) in this question's responses indicates that the living labs are to some extent 'children of the commission', for reasons discussed earlier, where the European Commission continues to support 'transnational activities' between European organisations who have already formed partnerships through R&D&I activities.

The living labs were then asked to say, for their region, how committed they would say organisations including 'Government, Other Public Sector Organisations, Universities, Private Sector Organisations, European Commission and Charities within their area' were to the concept of living labs. It was interesting to note that 33.9% of respondents reported Universities highest as 'Very committed' with the European Commission scoring second with 26.8%. About half of respondents said that most organisations were 'Somewhat committed'.

In a closely related question, the respondents were asked if their living lab had found it easy or difficult to develop relationships with 'Government, Other Public Sector Organisations, Universities, Private Sector Organisations and European Commission'. It was interesting, albeit perhaps not so unexpected, that 23.2% of respondents answered that they had not tried to develop relationships with government. This response reinforces the analysis that living labs are a regional phenomenon and many national governments (there are exceptions) are quite disconnected from living labs and do not have well-developed policy frameworks relating to living labs, or indeed, arguably more broadly to support user or citizen participation.

A significant minority of respondents, ranging from a quarter to a third, said it was difficult to develop relationships across all the organisations. However, a clear majority answered that across all organisations, they found it easy to develop relationships. This confirms that living labs generally have relationships across the triple-helix mix of organisations and have found it easy to develop these relationships.

4.9 Sustainability

The survey asked respondents how sustainable they considered their living lab to be. A clear majority of respondents (78.6%) believed their living lab to be sustainable in the short-term of 1-2 years. Over a medium time horizon of 2-5 years, this percentage fell to 57.1%. The fall can be attributed to growing uncertainty over time as the percentage of those who didn't know how sustainable their living lab would be over 2-5 years rose to 28.6% (from 5.4% not knowing about sustainability in the short term of 1-2 years). Over the longer-term period of 5 years, most people simply didn't know how sustainable their lab would be (76.8%). This increase in uncertainty over longer time periods may reflect the relatively precarious position of living labs, that while many are 'children of the Commission', their position is not underpinned by national legislation, their governance is a partnership of different interests and they are often regional actors with a regional remit and outlook.

The survey then asked the living labs what they believed were their top three challenges in the year to come. Many living labs provided the same three challenges but in all possible variations. These top challenges were, in descending order of priority: 1) funding; 2) getting more partners and end users (e.g., 'Get more external customers outside the region'); and 3) expanding activities and embedding user-centric activities in partners (e.g., 'Grow - to be able to manage more project in parallel').

4.10 Relating to Other Living Labs

In terms of their relationships with other living labs, respondents were asked with how many other living labs their living lab had connections. Just over half of respondents (51.8%) said that they had connections with four or more living labs with 26.8% saying that they had links with 1-3 living labs. The general degree of connectedness of living labs must be welcomed. However, It was revealing that 19.6% said that they had no connections with other living labs. On examination of the data, this group of living labs cited 'communication' as a challenge over the next year, and one of the living labs that had no connections with other living labs said that they worked with over 100,000 users.

In terms of how often living labs are in contact with other living labs, the responses indicate that the most common frequency is quarterly (43.2%) with the remainder split between less frequently than that (20.5%) and more frequently with 29.5% saying at least monthly and 6.8% saying at least weekly.

Those living labs that had four or more connections with other living labs had generally more frequent contact with those living labs, indicating that membership of a network brought with it more frequent interactions as a matter of course, perhaps related to the common 'network effect'.

4.11 Financial Support

The survey asked the respondents if access to funding had been a problem for their living lab and 83.9% said that it was a problem, with 25% saying it was a minor problem and 58.9% saying that it was a major problem. The reason for the uncertainty in the longer time horizon of 3-5 years and beyond 5 years evident in the responses earlier is perhaps now revealed to be access to funding and therefore the key issue in the future for living labs will relate to sustainability.

The living labs were asked a final question about their main sources of funding. The public sector accounted for 42.9% of funding sources, breaking down to 25% for government and 17.9% for other public sector organisations. Universities accounted for 14.3% of funding with private organisations contributing 10.7%. The European Commission accounted for 19.6% of funding, perhaps representing the support inherent in R&D&I activities from the instruments in the Framework 7 Programme (FP7) and the Competitiveness & Innovation Programme (CIP).

5 Discussion

Most living labs provided support for product and/or service development, primarily related to using new technologies. While living labs began as an urban phenomenon, almost two-thirds of living labs are now 'territorial', that is, they primarily operate at a regional level. Responses indicate that universities and private sector organisations are well embedded in the activities of living labs and it was interesting to note that a third of respondents reported universities being very committed to their living lab with the European Commission scoring second with a quarter. It was also interesting, albeit perhaps not so unexpected, that a quarter of respondents answered that they had not tried to develop relationships with national government. This response supports the findings that living labs are a regional phenomenon and many national governments are quite disconnected from living labs and do not have well-developed policy frameworks relating to living labs, or indeed, arguably more broadly to support user or citizen participation.

In terms of user numbers, many living labs involve small numbers of users but the majority support over 1,000 users. Responses indicate that superficial interaction with end users is relatively easy to do while more involved or complex interactions are somewhat more difficult. A clear majority of living labs answered that the translation process from end users to products or service change was difficult. It was interesting to note that some respondents found that the reason was a kind of 'lost in translation' effect, for example: "it can be hard to get constructive and instructive comments from users that are not used to giving feedback".

A majority of living labs would like access to support for evaluation and research training to assist with engaging with users and evaluating user involvement. Most living labs are inter-connected in some way with at least four other labs and communicated at least quarterly with the other labs with the main reason for this networking being to share experiences and knowledge.

The top challenges to living labs were given as funding; getting more partners and/or end users; and expanding activities and embedding user-centric activities in partners. The respondents indicated that funding their activities was a problem with most funding being project-based sourced primarily from non-private sector sources including public and academia. Most living labs simply didn't know how sustainable their lab would be over long time periods. This may reflect the relatively precarious position of living labs, that while many benefit from the European Commission's support for organisations to form partnerships through R&D&I funded activities, their position is not underpinned by national legislation, their governance is a partnership of differing and sometime competing interests and they are often regional actors with a regional remit and outlook with all that this entails.

What is remarkable about the findings is the diversity of purpose and scope of the living labs surveyed. We find living labs to be alive and healthy in 2011, somewhat uncertain about the future but enthusiastic about the challenges ahead to be tackled. It is apparent that many have a particular niche in which they operate. Some labs are region-based, others focus on a particular product family for example, automotive design, while others seek to address particular societal needs in, for example, healthcare. However, the use of technology to engage and support users as early as possible in product and service development is the common denominator for all of them.

Acknowledgements. The authors would like to thank Eileen Beamish, Social Research Centre Ltd and Donal McDade, Social Market Research Ltd for their support in the design of the survey. We would also like to thank the UK Higher Education Innovation Fund (HEIF) for its support to TRAIL, and to ENoLL for helping to promote the survey and its findings.

References

Mitchell, W.J.: Me++: the cyborg self and the networked city. MIT Press, Cambridge (2003)

Mulvenna, M.D., Martin, S., McDade, D., Beamish, E., de Oliveira, A., Kivilehto, A.: TRAIL Living Labs Survey 2011: A survey of the ENOLL living labs, University of Ulster, 40 pages (2011) ISBN-978-1-85923-249-1

Leading Innovation through Knowledge Transfer to Social Enterprises in Northern Ireland

Eddie Friel and Kerry Patterson

University of Ulster, Cromore Road, Coleraine, Co Londonderry, BT52 1SA

Abstract. This paper describes the work carried out by the University of Ulster to transfer knowledge and expertise to social enterprises. It sets the Northern Ireland social economy sector in context and highlights the exemplary manner in which Ulster supports social enterprises in innovative ways to help the sector prosper. A number of examples are explored to demonstrate how knowledge is transferred to the sector.

1 Introduction

The University of Ulster is the largest university on the island of Ireland and incorporates four main campuses: Jordanstown, Belfast city centre, Coleraine and Magee. Its vision is to be a university with a national and international reputation for excellence, innovation and regional engagement and it has a major direct and indirect impact on the economy and community in Northern Ireland.

2 Background

In 2008, the Northern Ireland Executive Programme for Government identified "growing a dynamic, innovative economy" as the top priority over the next 3 years [1] with the Department of Enterprise, Trade and Investment (DETI) taking lead responsibility to deliver this aim. Within this context, DETI launched the Social Economy Enterprise Strategy 2010/11 [2] which continued the Executive's ongoing commitment to three strategic objectives to increase awareness of the sector, develop its business strength and provide a supportive environment in which it can prosper.

The social economy is a wide and diverse sector which has been operating and developing over many years and has a strong tradition of supporting local communities across Northern Ireland. The accepted definition of social enterprise across NI Government departments and The Third Sector office of the UK Cabinet Office states that a social enterprise must have a social, community or ethical purpose; a commercial business model and a legal form appropriate to a not-for personal-profit status.

R.J. Howlett et al. (Eds.): *Innovation through Knowledge Transfer 2012*, SIST 18, pp. 145–147.
DOI: 10.1007/978-3-642-34219-6_16

3 Ulster's Innovation Engagement

Ulster's Office of Innovation leads on the University's engagement with companies, social enterprises and community groups. Ulster recognises that it has a unique role to play in contributing to the growth and development of the social economy sector, particularly increasing the business strengths of social enterprises. In its Corporate Plan it sets "Contributing to the economic, social and cultural inclusion in the region" as a cross cutting aim.

4 Social Enterprise Support at Ulster

Historically social enterprise activity at Ulster was led by staff from Social Sciences, primarily through their engagement with the local community and voluntary sector. However it was decided to take a more strategic approach to support the social economy sector and the University's Business Institute which is at the leading edge in terms of curriculum and training materials led the engagement.

Since 2005, the University has provided targeted assistance with support from the Department for Employment and Learning's (DEL) HEIF funding to facilitate the sharing of Ulster's expertise with social enterprises. The University has recognised the need for social enterprises to be innovative and adopt good business practice to ensure their survival. Activity is very much in line with UK Government's efforts to challenge Universities to consider how they can encourage businesses to work with Higher Education establishments for the first time.

Much progress has been made within a short space of time with activity cited in the Progress on Developing a Successful Social Economy: NI Government's Progress Report on 2006-2007 Action Plan (July 2007) [3]. The University's work to support the sector has also been recognised in the annual Business in the Community Awards as reported by DEL in its review of the Higher Education Innovation Fund. [4]

5 Support Exemplars

Ulster plays a distinct role in promoting creativity and innovation within the social economy sector and offers a 'best in class' service to social enterprises. It provides bespoke training, as well as access to other relevant mainstream University support mechanisms including Innovation Vouchers and the Knowledge Transfer Partnership programme. Recognising that social enterprises often suffer from a lack of business skills and confidence to make them successful, Ulster's focus is on helping them to provide a competitive, innovative and client-focused service, which in turn strengthens their social impacts.

Specific interventions include a Knowledge Transfer Partnership with Bryson Charitable Group, the largest social enterprise in Northern Ireland. The KTP, which is now in its second year focuses on an issue at the heart of the sector; the challenge of being able to demonstrate and be recognised for social impact in public procurement. Bryson's KTP is developing end to end process value mapping to inform and influence public procurement guidance, processes and training and will develop a social enterprise "Best Practice" tendering process model which will incorporate cost analysis, social outcomes and qualitative criteria.

Ulster has also been successful in highlighting the relevance of the Invest Northern Ireland Innovation Voucher scheme, sharing its expertise with over 20 social enterprise projects. Onus (NI) who Ulster, having identified the need to carry out market research to ascertain the drivers and inhibitors that underpinned adoption of a workplace domestic violence policy, helped to develop a Communication strategy to launch a corporate charter on domestic violence; a project with EXTERN Recycle to examine safety issues relating to the recycling of LCD monitors; and a pilot study for the Ulster Cancer Foundation on breast cancer reconstruction following mastectomy to enhance patient communication and information using novel digital technologies. Projects about to get underway include support for Stepping Stones who employ adults with learning difficulties. Ulster's academics from both Design and Occupational Therapy will identify new products for the adults to make, ensuring that the level of expertise required for each product is appropriate to different learning challenges.

Support has also been given through a unique University programme, the Science Shop, which provides student research assistance to social enterprises. Compass Advocacy Network (CAN) who run a number of separate recycling social enterprises benefitted from a review carried out by students of their existing marketing strategy and 'brand'. The result was the creation of new branding connecting all of CAN's social enterprises while fully integrating their core values at the heart of their new marketing campaign.

In addition the University has also provided assistance by enabling a key member of staff to take up a Board position on the Regional Social Economy Network for the last three years. Overall the University's engagement with the sector has grown significantly following the intervention of the Office of Innovation.

6 Conclusions

The University of Ulster is at the forefront in Northern Ireland in terms of its impacts on social enterprises through teaching and learning, research and academic enterprise. Through a combination of indirect forms of engagement and a number of direct strategic intervention measures, the University has contributed towards the innovation and sustainability of the sector right across Northern Ireland.

References

1. Northern Ireland Executive Programme for Government 2008-2011, p. 8 (January 2008)
2. Social Economy Enterprise Strategy 2010-11 (March 2010)
3. Progress on Developing a Successful Social Economy: NI Government's Progress Report on 2006-2007 Action Plan, p. 3 (July 2007)
4. DEL and Invest NI Project Evaluation of the Second Round of the Northern Ireland Higher education innovation Fund (NI HEIF 2) Report, p. 85 (May 2010)

Planes, Trains and Automobiles: The Importance of Location for Knowledge Transfer in the Transportation Sector*

Jeffrey A. Moore[1], Daniel K.N. Johnson[2,**], and Kristina M. Lybecker[2]

[1] Colorado College's Department of Economics and Business, 14 East Cache La Poudre Street, Colorado Springs, CO 80903
jeffrey.moore@ColoradoCollege.edu
[2] Colorado College, 14 East Cache La Poudre Street, Colorado Springs, CO 80903
{djohnson,Kristina.Lybecker}@ColoradoCollege.edu

Abstract. Using over 200,000 U.S. patent citations, we test whether knowledge transfers in the transportation sector are sensitive to distance, and whether that sensitivity has changed over time. Controlling for self-citation by inventor, assignee and examiner, multivariate regression analysis shows that physical distance is becoming less important for spillovers with time, albeit in a nonlinear fashion.

Keywords: transportation, patent, citation, spillover, distance.

Historically, firms within an industry have often clustered geographically, due to localization economies (Henderson (1986) or Smith and Florida (1994)) such as the speed (see Caballero and Jaffe (1993) and tacitness of the transfer of knowledge (see Von Hippel (1994)). To our knowledge, however, this phenomenon has not been measured within the transportation sector, nor has its importance been compared over time for this sector.

This paper examines knowledge flows within the transportation industry, both confirming traditional evidence that inter-firm knowledge transfers typically decreased with distance, and documenting their increase over time (albeit at a decreasing rate). Figure 1 portrays the pattern without any statistical correction for other potentially correlating factors.

* Special thanks to a team led by Nicole Gurley for their patience while geo-coding, to Matt Gottfried for his continued patience with our GIS-based questions, and generous support from Colorado College in the forms of the Schlessman Professorship, a Faculty Grant, a Mrachek Research Fellowship, a Mellon Research Block and assistance from Colorado College's Chapman Fund.

** Corresponding author.

R.J. Howlett et al. (Eds.): *Innovation through Knowledge Transfer 2012*, SIST 18, pp. 149–157.
DOI: 10.1007/978-3-642-34219-6_17

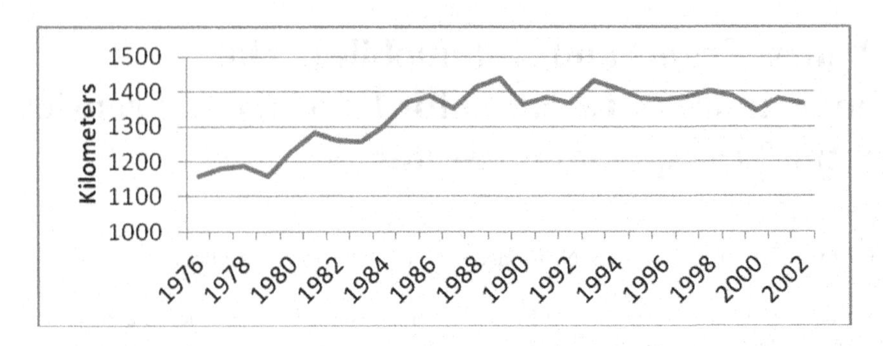

Fig. 1 Average citation distance (in kilometers)

In section 2 of the paper, we briefly review the relevant literature on the geographic nature of knowledge spillovers. Section 3 describes our data set, designed for compatibility with the literature, and Section 4 presents multivariate regression analysis that controls for non-geographic effects in presenting the declining role of distance. Section 5 concludes with implications for policy and further research.

2 Literature Review

Most of the literature suggests that knowledge spillovers cluster geographically, with higher spillovers (shown by more patent citations) within a short distance. The underlying supposition is that inventors are more aware of (or find more use for) inventions located close to them, and therefore build more heavily on local inventions. Alternatively, it might be a function of where innovations actually occur, smaller distances if the industry clusters around a central hub. The result is a geographic clustering of citations.

Empirical evidence stresses the role of geography in the transfer of knowledge within an innovation network (see, for example, a review by Gelsing, 1992), and emphasizes the importance of frequent personal contact and research collaboration. Lundvall (1992) points out that the importance of geography should differ predictably by technology type. While geography has little impact on stationary technologies (which face constant needs and opportunities), its importance grows quickly for technologies undergoing incremental, and particularly, radical innovation.

Geographic proximity has already been used to explain the location choices of R&D-intensive activities (e.g. Dorfman (1988)). However, the location of *firms* is not always a good predictor of the location of *innovation* (Feldman (1994), Johnson and Brown, (2004)).

Localization of patent citations has been firmly established (Jaffe et al., 1993), with a random sample of patents clearly more likely to cite local patents than patents by parties that are located farther away, at every geographically aggregated level. The effects are more intense where knowledge becomes obsolescent rapidly, such as electronics, optics, and nuclear technology (Jaffe and Trajtenberg, 1996). Alternatively, strong voices in the literature argue that either distance has never mattered much (Thompson and Fox-Kean, 2005), or that the impact of communication technology on productivity or on knowledge transmission across distance will not be that great (Vasileiadou and Vliegenthart, 2009; Graham, 2001).

3 Data

3.1 Measurement Issues

This paper relies exclusively on patent citations from transportation-sector patents as a geographic measure of knowledge spillovers in the sector. When a patent application is submitted for approval, it is accompanied by a list of citations to other patents and literature which have been instrumental in the creation of this technology, or which delineate the legal limits of this application. The intention is twofold: to build a convincing case that this application is novel and unobvious to someone trained in the field, and to provide a legal record in order to protect patent rights in the future. To this list of citations, a patent examiner may suggest his or her own list of citations for the applicant to include. The result is a paper trail of knowledge creation.

Of course, patents records do not perfectly reflect the creation of technology, as some innovations are never patented and patents vary greatly in size and importance (Feldman (1994)). Likewise, citations do not perfectly reflect the transfer of knowledge, as they may be inserted for a variety of other reasons including legal protection or examiner privilege (Jaffe et al. (2000)). However, their statistical tests indicate that overall citations can be interpreted as a signal of spillovers, albeit a noisy signal.

In order to define the scope of the study, we follow the World Intellectual Property Organization's definition for transportation, definitions which include portions of 5 separate International Patent Classification (IPC) category codes on the 4-digit level: B60 through B64 inclusive. We appended our dataset with the set of patents cited by transportation patents, at least those that were themselves granted between 1976 and 2002. While the citing patents date exclusively from this period, patents cited by our observation sample may predate 1976, but were truncated from consideration simply due to data availability. Furthermore, citing and cited patents from all non-U.S. inventors have been excluded, for reasons of feasibility. However, Johnson and Sneed (2009) present evidence in the literature that international citations are increasing in frequency across a host of technologies, evidence which is sympathetic to the hypothesis here that citation distances have been increasing over time (Johnson and Lybecker, 2011).

Our focus is thus exclusively upon transportation-sector knowledge flows within the United States, within a banded period of time, inviting subsequent scholars to continue to work of expanding the dataset's coverage.

3.2 Clustering of Knowledge Citations

Patent citations may cluster for non-geographic reasons, coincidentally causing a pattern that appears geographic merely through correlation with other phenomena. For example, inventors may be more familiar with their own patents, and the same may be true of assignees (the legal holders of the intellectual property rights) if employees of a firm are familiar with other patents held by the same firm. Therefore we include self-citations in the analysis but control for them separately.

Using U.S. patent data from a combination of sources (NBER website as described in Hall et al., 2001, in addition to raw data collected by the independent firm MicroPatent), this paper relies on citations collected from all transportation-sector patents granted between 1976 and 2002. Each citation's endpoints (citing and cited) were then geo-coded for the primary location of each U.S.-based patent at the geographic center of the city listed (as specific addresses are available for less than ten percent of all patent documents).

The result is a dataset of 203,158 U.S.-based transportation patent documents that include a total of 263,796 citations to other U.S.-based patent documents. Previous literature (e.g. Johnson and Lybecker, 2011) indicates that each of the following factors may play some role in the distance of a citation, so this research measured each for every observed citation:

- whether patents k and K have the same inventor (SI),
- whether patents k and K have the same assignee (SA),
- whether patents k and K are in the same technology cluster (ST),
- how similar the citing and cited states are in technology types (SC),
- whether the cited patent is also classified as transportation (T),
- whether the assignee is a government agency (G),
- whether the assignee is an educational institution (U),
- how old the citation is, in years between citing and cited patent (A), along with its squared term to account for the potentially nonlinear effects of age (A^2), and
- year of citing patent K, to account for citation inflation (Y).

Self-citation by inventors accounted for almost three percent of all citations from transportation patents, a much lower self-citation rate than has been documented in other sectors like biotechnology (twelve percent, according to Johnson and Lybecker, 2011). On the other hand, thirty-four percent of all citations were to the same assignee, higher than has been measured in biotechnology (twenty-four percent, according to Johnson and Lybecker, 2011).

It is also possible that patents closer in technology space may have longer or shorter citation distances than more diverse cited patents. The data are coded so that a binary variable, ST, indicates whether the International Patent Classification (IPC) system places both citing and cited patents in the same technology cluster, out of 634 mutually exclusive and exhaustive clusters in the system. This system, in global use since 1975, is the standard by which all patents are organized (and thus assigned to examiners for processing, or searched by inventors and lawyers to establish claims).

The fourth variable, the technological correlation between citing and cited states (SC), is included for a similar reason. Each state's technological profile was calculated as the share of patent activity assigned to each of the 634 IPC technology classes. Pair-wise correlations between state vectors then provide a measure of technological similarity between locations. The analysis also includes an indicator of whether the cited patent is classified as transportation.

Because government (G) and university (U) patents may cite knowledge differently than do private sector patents, we include those indicators as controls as well. Linear and squared age terms are included to accommodate nonlinear effects for older knowledge. Finally, since the goal of the analysis is to test whether distance changes over time, it is necessary to include dummy variables for each time period (year Y).

Summary statistics of the 203,158 citations in the sample are presented in Table 1. The average citation is around 1375 kilometers long, but the distribution has a very wide variance. For this reason, tests are performed not only using distance as a dependent variable, but on alternative specifications using a dependent variable of the logarithm of distance.

Table 1 Summary statistics for citation dataset

Variable	**Mean**	**Std. Dev.**	**Min**	**Max**
Citation distance (km)	1375.48	1272.63	0	12565.15
SC= correlation in technology portfolios between cited and citing states	0.84	0.15	0.01	1
A= age of citation (years)	8.38	5.91	0	27
	Number of zeros			
SI = same inventor in cited and citing patents	196846 (97.0%)		0	1
SA = same assignee in cited and citing patents	133889 (66.0%)		0	1
ST = same IPC in cited and citing patents	98900 (48.7%)		0	1
T = cited patent is also Transportation	63470 (31.2%)		0	1
G = government assignee	200,595 (98.7%)		0	1
U = university assignee	202,030 (99.4%)		0	1

4 Model and Results

Our regression analysis follows the literature (Johnson and Lybecker, 2011) in using a simple model by Petersen and Rajan (2002) with the patent citation as the unit of analysis. The model expresses the distance between a cited patent *k* granted in year *y* and a subsequent citing patent, *K*, granted in year *Y*, as a function of the attributes of patents *k* and *K:*

$$\delta_{k,K} = \alpha(k,K) + \varepsilon \tag{1}$$

where and $\delta_{k,K}$ represents the distance between patents *k* and *K*, $\alpha(k,K)$ is a vector of the other attributes of patents *k* and *K* that affect the probability of citation, and ε is a randomly distributed error term. We postulate a reduced functional form, using the log of [distance plus one] in order to avoid taking the log of a zero distance) because the fit of the equation is better:

$$\begin{aligned} &\text{Distance} = \alpha_0 + \alpha_{\text{SameAssignee}}\text{SA} + \\ &\alpha_{\text{SameInventor}}\text{SI} + \alpha_{\text{SameTech}}\text{ST} + \\ &\alpha_{\text{Transportation}}\text{T} + \alpha_{\text{Govt}}\text{G} + \alpha_{\text{University}}\text{U} + \\ &\alpha_{\text{Age}}\text{A} + \alpha_{\text{AgeSquared}}\text{A}^2 + \textstyle\sum_{1976}^{2002} \alpha_{\text{Year}=i}\text{Y}_i + \\ &\varepsilon_K + u \end{aligned} \tag{2}$$

We include a fixed effect specific to the citing patent (ε_K), since there are presumably immeasurable factors which might dictate a longer or shorter average citation distance.

Table 2 presents multivariate regression Tobit estimates (left-censored for intra-city citations with a distance of 0 kilometers), with White-corrected errors, using fixed effects. Considering all citations, the average distance unambiguously increases with time, with strong evidence of a non-linear pattern. Sensitivity tests find very similar results if we restrict our consideration to citations of more than 10 kilometers, more than 50 kilometers, or more than 100 kilometers. Results for citations spanning more than 100 kilometers (i.e., excluding short and intra-city citations) are presented in the second panel of Table 2, and tell a similar story.

To permit maximum flexibility to these nonlinearities, and potential nuances in particular years, we performed the same analyses using separate year indicator variables instead of a time trend variable.

Unsurprisingly, citations with the same assignee or same inventor are more likely to be proximate than are other citations, an effect is especially strong and significant for inventors. Citations within the same technology class are closer to each other than more dissimilar patents (the ST coefficient is negative), and states that have similar technology portfolios tend to be close together, a fact captured by the negative coefficient on that variable (SC).

Table 2 Tobit weighted regressions on log(distance+1)

	All citations		**Only citations with distance>100km**	
variable	**coefficient**	**t-statistic**	**coefficient**	**t-statistic**
SA	-0.41	(-30.08)***	0.18	(26.51)***
SI	-3.57	(-66.99)***	-0.59	(-12.69)***
ST	-0.29	(-18.26)***	-0.06	(-6.61)***
SC	-6.36	(-130.57)***	-1.33	(-56.03)***
T	0.14	(8.11)***	0.06	(6.5)***
G	0.16	(3.41)***	0.23	(8.53)***
U	0.34	(5.04)***	0.22	(5.63)***
A	$6.96\text{x}10^{-2}$	(19.46)***	$8.60\text{x}10^{-3}$	(4.54)***
A^2	$-1.98\ \text{x}10^{-3}$	(-13.42)***	$-2.86\text{x}10^{-4}$	(-3.56)***
Y	$1.23\ \text{x}10^{-2}$	2(.49)**	$1.58\text{x}10^{-2}$	(6.14)***
Y^2	$-3.50\ \text{x}10^{-4}$	(-2.42)**	$-4.35\text{x}10^{-4}$	(-5.83)***
constant	11.38	(212.22)***	7.92	(275.75)***
F-stat		(2583.47)***		(418.31)***
Observations		203158		165647

Notes: *** indicates 99% confidence, ** 95% confidence, * 90% confidence. Implicit impacts are calculated at the sample mean for the group in question.

Citations that cite other transportation patents average a slightly longer distance than their peers. Apparently distance matters less for the transfer of purely transportation-related knowledge than it matters for the transfer of non-transportation innovations into the transportation sector.

Government-assigned knowledge tends to travel longer transmission distances for the knowledge they cite, a result that is more pronounced when we consider only long-distance (>100 km) citations. Academic patents tend to be longer than their peers as well, but that effect is less pronounced among long-distance citations (>100 km). Finally, age matters; older citations travel longer distances, an effect which other studies (e.g. Johnson and Popp, 2003) have confirmed.

The results point to the fact that physical distance has become less of a constraint with the passage of time. Perhaps the trend is due to the nature of the knowledge being created, but we suspect that it is more due to advances in communication, which allows easier transmission of information across great distances (e.g. Kim et al., 2006). In short, the principles underlying the inter-firm transfer of knowledge are changing in a striking fashion, making spillovers easier and longer than ever before.

5 Conclusion and Policy Implications

We are left with a striking picture of the inter-firm transfer of transportation technology knowledge. Controlling for other factors, knowledge flows historically diminished with physical distance, but the importance of distance has been receding with time. That is, knowledge is more likely to transfer over longer distances now than it was twenty years ago. Long-distance knowledge transfers are increasingly the norm in transportation technology. Innovation has become possible at a wider array of locations, potentially drawing on a wider range of raw materials and ideas. This might imply a potential for the deliberate fostering of non-traditional locations for transportation technology, with a prerequisite of vibrant communication with the research community elsewhere.

References

Caballero, R.J., Jaffe, A.B.: How High are the Giants' Shoulders: An Empirical Assessment of Knowledge Spill-overs and Creative Destruction in a Model of Economic Growth. In: Blanchard, O.J., Fischer, S. (eds.) NBER Macroeconomics Annual 1993. MIT Press, Cambridge (1993)

Dorfman, N.S.: Route 128: The Development of a Regional High Technology Economy. In: Lampe, D. (ed.) The Massachusetts Miracle: High Technology and Economic Revitalization. MIT Press, Cambridge (1988)

Feldman, M.P.: The Geography of Innovation. Kluwer Academic Publishers, Dordrecht (1994)

Gelsing, L.: Innovation and Development of Industrial Networks. In: Lundvall, B. (ed.) National Systems of Innovation: Towards a Theory of Innovation and Interactive Learning. Pinter Publishers, London (1992)

Graham, A.: The Assessment: Economics of the Internet. Oxford Review of Economic Policy 17(2), 145–158 (2001)

Hall, B.H., Jaffe, A.B., Trajtenberg, M.: The NBER Patent Citations Data File: Lessons, Insights and Methodological Tools. NBER Working Paper #8498 (2001)

Henderson, J.V.: Efficiency of Resource Usage and City Size. Journal of Urban Economics 19, 47–70 (1986)

Jaffe, A.B., Trajtenberg, M., Henderson, R.: Geographic Localization of Knowledge Spillovers as Evidenced by Patent Citations. Quarterly Journal of Economics 108, 577–598 (1993)

Jaffe, A.B., Trajtenberg, M.: Flows of Knowledge From Universities and Federal Labs: Modeling the Flow of Patent Citations Over Time and Across Institutional and Geographic Boundaries. NBER Working Paper #5712 (1996)

Jaffe, A.B., Trajtenberg, M., Fogarty, M.S.: Knowledge Spillovers and Patent Citations: Evidence from a Survey of Inventors. American Economic Review 90, 215–218 (2000)

Johnson, D.K.N., Brown, A.: How the West Has Won: Regional and Industrial Inversion in U.S. Patent Activity. Economic Geography 80(3), 241–260 (2004)

Johnson, D.K.N., Popp, D.C.: Forced Out of the Closet: The Impact of the American Inventors Protection Act on the Timing of Patent Disclosure. Rand Journal of Economics 34(1), 96–112 (2003)

Johnson, D.K.N., Lybecker, K.M.: Does distance matter less now? The changing role of geography in biotechnological innovation. Review of Industrial Organization 40(1), 21–35 (2011)

Johnson, D.K.N., Sneed, K.A.: Are Many Heads Better Than Two? Recent Changes in International Technological Collaboration. Journal of Law, Ethics and Intellectual Property 3(1), 1–9 (2009)

Kim, E.H., Morse, A., Zingales, L.: Are Elite Universities Losing their Competitive Edge? CEPR Discussion Paper 5700 (June 2006)

Lundvall, B.A.: User-Producer Relationships, National Systems of Innovation and Internationalisation. In: Lundvall, B. (ed.) National Systems of Innovation: Towards a Theory of Innovation and Interactive Learning. Pinter Publishers, London (1992)

Petersen, M.A., Rajan, R.G.: Does Distance Still Matter? The Information Revolution in Small Business Lending. The Journal of Finance 57(6), 2533–2570 (2002)

Smith, D., Florida, R.: Agglomeration and Industry Location: An Econometric Analysis of Japanese-Affiliated Manufacturing Establishments in Automotive Related Industries. Journal of Urban Economics 36, 23–41 (1994)

Thompson, P., Fox-Kean, M.: Patent Citations and the Geography of Knowledge Spillovers: A reassessment. American Economic Review 95, 450–460

Vasileiadou, E., Vliegenthart, R.: Research Productivity in the Era of the Internet Revisited (2009), http://ssrn.com/abstract=1431457

Von Hippel, E.: Sticky Information and the Locus of Problem-Solving. Management Science 40, 429–439 (1994)

Information Sharing among Innovative SME: An Exploratory Study within the Portuguese SME Innovation Network

Esther Lage and Bráulio Alturas

ISCTE – Instituto Universitário de Lisboa (University Institute of Lisbon)
{elleo,braulio.alturas}@iscte.pt

Abstract. This paper aims to enhance understanding on inter-organizational information sharing through a study in a Portuguese SME Innovation Network, which is composed of innovative firms from different economical sectors. The type of information shared was identified as well as the channels used, the gains obtained and the factors that influence its occurrence. A qualitative approach was used based on the perception of members and coordination. The motivating role played by the coordination of the network is a point to emphasize regarding the collective sharing of information. Factors related to network structure, national culture, characteristics of the companies and channels used were identified as inhibitors to a greater identification between the companies. Moreover, the factors that influence the information sharing process between the companies and its most important information sharing partners within the network were also identified.

1 Introduction

Despite the growing number of studies on inter-organizational information/ knowledge sharing in the last 20 years (Easterby-Smith et al. 2008), there are still many aspects that require a deeper understanding, such as the treatment of information sharing as an independent variable from other factors that define collaboration or cooperation (Madlberger 2009) and the identification of its antecedents, especially from a multi-disciplinary perspective (Wijk et al. 2008; Martinkenaite 2011;).

This paper aims to enhance the understanding on inter-organizational information sharing through a study in a Portuguese SME Innovation Network, which is composed of innovative firms from different economical sectors. The type of information shared within the network will be presented as well as the channels used, the aspects influencing the process and its contribution to the organizations. These aspects were identified both in the collective scope and between the most important relationships. A qualitative approach was used based on the perception of members and coordination.

Understanding the kind of information shared, how and why information flows among partners will bring gains to the network coordinators, who are expected to encourage the improvement of policies, incentives and channels concerning the in-

R.J. Howlett et al. (Eds.): *Innovation through Knowledge Transfer 2012*, SIST 18, pp. 159–168.
DOI: 10.1007/978-3-642-34219-6_18

formation sharing process. Moreover, the results may offer guidance to the companies when it comes to the aspects that must be developed internally or stimulated in the partner when the information sharing is aimed.

2 Information Sharing in Inter-organizational Networks

Inter-organizational networks can be understood as "institutional arrangements that allow efficient organization of economic activities through the coordination of systematic links established among interdependent firms" (Britto 2001). One of the links that characterize any kinds of networks is the sharing of information.

There seems to be no consensus regarding the most appropriate term to refer to the inter-organizational information sharing, since various terms are used in the literature such as *information/knowledge sharing* (Fiala 2004; Mei and Nie 2007), *information dissemination* (Zhang, Vonderembse and Lim 2006), *information/knowledge transfer* (Tushman and Scanlan 1981; Simonin 2004), *information exchange*, (Moberg et.al. 2002; Madlberger 2009), *knowledge flow* (Dahl and Pedersen 2004), *knowledge diffusion* (Ernst and Kim 2002), *knowledge acquisition* (Hau and Evangelista 2007). In this paper information sharing is defined as the information's communication process amid the members of the network. This process is understood to take place in a social context, which means that the informational needs of individuals are not merely cognitive, but directly related to labour and social groups to which they belong (Capurro 2003). Information is defined as meaningful data.

Researches conducted in different types of inter-organizational relationships, such as supply chain networks (Moberg 2002; Madlberger 2009), strategic alliances (Simonin 2004), clusters (Mei and Nie 2007) and innovation networks and inter-organizational teams (Fritsch and Kauffeld-Monz 2008; Bond III et al. Tang 2008) have showed that the information sharing process is dynamic and its occurrence requires taking into consideration several factors, such as the type of partnership, the type of information, the characteristics of the firms and the characteristics of the network.

Specifically apropos its antecedents, several factors are identified in the literature, such as aspects related to the firm's characteristics (organization size, absorptive capacity, ability to transfer, motivation to teach and to learn, a centralized position in the network, the presence of ICT), to the nature of the knowledge (tacitness, complexity, specificity, ambiguity) and to the inter-organizational dynamics (power, trust, social ties, the existence of vision and common systems and exchange routines) (Wijk et al. 2008; Easterby-Smith et. al. 2008; Madlberger 2009; Martinkenaite 2011).

Finally, literature highlights the benefits of inter-organizational information sharing, such as prerequisites for the consolidation of inter-organizational relations, antecedents for the improvement of product development (Lawson et al. 2009); antecedents for the innovation capacity and performance of organizations (Mei and Nie 2007; Wijk, Jansen and Lyles 2008); antecedents for the development of new strategies and customer satisfaction (Zhang et al. 2006).

3 Methodology

Taking into account the assumptions underlying the qualitative study according to Creswell (1994), this study can be classified as a qualitative research. The focus is on the *process* of information sharing, mainly considering the influencing factors and channels used; the main interest is the *meaning* given by the coordination representative and the companies' managers about the process; the result is a *descriptive* understanding of the process and an *inductive* analysis of the data.

When it comes to the objective it is an exploratory research and when it comes to the method it is classified as a field research. The unit of analysis was the information sharing process within the "SME Innovation COTEC Network", which belongs to the COTEC Portugal Network. Primary data was collected through semi-structured interviews with a representative of the network's coordination and the managers of five companies that belong to the network, as following:

- Interviewee A – Executive President – Pharmaceutical sector
- Interviewee B – Managing Partner – ICT sector
- Interviewee C – General Manager – Environmental Consulting sector
- Interviewee D – Chairman of the Board - ICT and engineering sector
- Interviewee E – Managing Director – Aerospace sector

Considering the exploratory feature of the study, a more diverse sample was aimed in order to obtain a broader view of the information sharing process. Therefore the five companies surveyed belong to different sectors. The time of foundation is situated between 10 and 22 years and the number of employees between 13 and 180. The companies are sited in different locations. Secondary data from the network website was also used. An interaction between the theoretical literature and the primary and secondary data was carried out, following Yin's (2009) orientation.

4 COTEC Portugal Network and SME Innovation COTEC Network

4.1 *Characterization of the Network*

COTEC Portugal Network - Business Association for Innovation - is a private, nonprofit organization, founded in 2003 by the President of the Republic, who has exerted, since the beginning, the role of President of the General Assembly. It aims to contribute to the National System of Innovation by promoting the competitiveness of Portuguese companies through the development of a culture and practices focused on innovation. The network was initially formed by large companies, called "Associated Companies".

In 2005, COTEC Portugal Network, taking into consideration the importance of the SME for the Portuguese economy, created the SME Innovation COTEC Network with the objective of promoting public recognition of innovative SME, establishing a cooperation between the Associated Companies and the SMEs and supporting SMEs growth phases especially concerning the attraction of investment and internationalization. SMEs are considered companies with less than 250 employees and annual turnover not exceeding 50 million or balance value not exceeding 43 million Euros.

The SME Innovation COTEC Network is composed of firms with a turnover of at least 200,000 Euros, at least 10 employees, at least 3 years in the market and a minimal degree of innovation. This degree is measured by a tool developed for this purpose - the innovation scoring - whose results are analyzed by a specific committee. Membership requires the payment of an annual fee of 1,000 Euros. Moreover, in order to remain in the network, companies must maintain the efforts towards innovation, since they will be evaluated annually with respect to this criterion. Nowadays the network has 119 members. According to the coordination' representative, around 40% of them belong to ICT sector and the remaining belongs to different sectors such as food, footwear, mold, engineering and furniture. The companies are also geographically dispersed over the country.

4.2 Information Sharing Process via Collective Channels

Two types of information are primordially shared within the network through the collective channels: information related to the companies - Associates or SMEs - and information about innovation. The main channels used by the SME Innovation Network to share information are the following ones:

- COTEC Network Website, which provides information such as activities promoted by the network to support the innovation process, training, information on innovation in Portugal, statistics on R&D and management practices.
- Technology Platform "Collaborate", a virtual communication channel that links the Associates and the SME, through which it is possible to share insights and launch projects.
- "Day of the Associate", on which representatives of the companies belonging to the SME Innovation Network are received by a large company that will share information on its competences, structure, expectations, experiences and buying trends, among others.
- The SME Innovation COTEC annual meeting, in which the members of the network have the opportunity to share successful experiences of other companies, especially regarding the conquest of new markets. The coordination has the purpose of maximizing the sharing with respect not only to successful experiences, but also to failures or mistakes, giving opportunities for companies to learn from the mistakes of each other.

The representative of the coordination highlights the important role played by coordination to encourage members to participate in network activities and to meet one another:

> We are the engine, we pull things. We try to make them go…and then we get out of the way as soon as possible.

He also highlights the importance of a sense of reciprocity among the members, a perception that things are ambivalent, as a condition that facilitates the information sharing.

Three gains are identified regarding the collective sharing of information. The first one, which comes especially from the information sent by the coordination of the Network, is the encouragement and training of members in respect of internal organization and management of innovation, as showed by the following statement:

> …helping companies to systematize the ideas, to organize and manage them, an aspect that is fundamental and which was managed by us empirically and based on feeling... (Interviewee A)

Interviewee A highlighted that the questionnaire itself, made by COTEC to select companies that want to belong to the network, help them in the systematization of ideas on innovation. Hiring people to fill positions such as manager of innovation and director of innovation, research and development and obtaining certification such as the standard 4457 on Research, Development and Innovation Management Systems are some of the strategies adopted by companies to improve the internal organization.

The second gain is the development of closer relationships between partners due to joint participation in the same events, which allows richer information sharing between them and identification of new business opportunities.

The third gain is the formation of new partnerships between network members. Interviewee B highlights a new partnership that was identified in a networking lunch. However, this gain is seemingly partially achieved. Only in the case of interviewee B the most important information sharing relationships started after the participation in COTEC. Nevertheless, the respondent points out that this was not a decisive factor, just one more element among other factors that reinforce the positive recognition of the company in the market.

Five inhibitors regarding the development of closer relationships among the companies and further information sharing were identified. The first one is related to the organization of events. Interviewees B and E feel the need for a more fruitful and closer exchange of information among the companies, a communication process that allows for a greater number of actors to speak in a closer or more informal way:

> I think communication is not the deepest, most effective or prolific, I think it is still largely confined to the events that are organized… (Interviewee *B*).

A second aspect refers to the absence of concrete proposals that would allow the grouping of companies with related interests and skills. According to respondent D, this is a time consuming task, since it requires the identification of potential customers / donors, funding sources, idea generation, consortia building and proposals generation. The absence of these proposals may indicate the companies' preference for working in consortia with companies outside the network, for working alone or a lack of resources such as time, money and expertise to develop such proposals. It is interesting to notice that, considering these two first aspects, both the more informal communication and structured mechanisms in the form of consortia are considered important to the members as channels to improve the identification of common interests.

The third aspect identified is the heterogeneous composition of the network. Respondent B acknowledged not seeking information on an intentional or systematic way because his company's sector is not well represented in the network, meaning that it is difficult to find some complementarities. According to Fritisch and Kauffeld-Monz (2008), who found no relationship between the heterogeneity of skills and information sharing in innovation networks in Germany, this fact could indicate a greater interest by the companies for exploitation than for exploration of new ideas.

The fourth aspect is related to national culture. According to interviewee D, the absence of a culture of collaboration in Portugal has been a hindrance to formation of partnerships. It is also, according to the respondent, an aspect that the network has been unable to overcome:

> It is not exactly a great culture for gathering or collaborating ... apart from some examples, each one is in his/her house doing what they think they are good at. And this extends to the business environment.

However, interviewee B, without referring specifically to the Portuguese culture, emphasized the importance of maintaining partnerships:

> ... We also understand that it is better trying to cultivate partnerships than being here alone in our corner; we always have a more limited view of reality.

In this matter, Moreira (2007, p.28) considers that the Portuguese culture is characterized by the absence of a collective system of trust and the predominance of strong informal neighboring relationships. The consequence of that is a cultural understanding of trust as the sum of individual parts and not as something that benefits the whole. This understanding inhibits inter-organizational, formal and organized cooperation and stimulates spontaneous, informal and clandestine ones.

Finally but not least, the commitment of top management to the network was identified as a relevant factor. Some respondents reported lack of time and resources to take part in the network's activities and focus on other priorities such as the relationship with international clients. This aspect becomes more relevant if one takes into account the cultural feature highlighted above and the important role played by the managers in small and medium companies.

4.3 Information Sharing Process via the Most Important Relationships

Beyond the comprehension of information sharing under the collective dimension, its process between the interviewed companies and their most important information sharing partners was also analyzed. Regarding the number of most important information sharing partners and the type of relationship, interviewee C mentioned 3 clients; interviewee D mentioned 3 partners in common projects;

interviewee B mentioned both, 6 clients and 1 partner in common project; and interviewee E mentioned 4 companies from the same sector with whom he keeps, nowadays, no business relations. Only interviewee A mentioned not having any relationships with other companies within the network.

When it comes to size, there are relationships with large companies and SME. Most of the clients are large companies. All partners in joint projects are SME. Respondent D related that although his company has projects in common with two larger companies than yours and one smaller, this aspect is not significant in terms of their importance. Despite the size, each one stands out for a specific competence. Respondent E highlights the aspect of competition between companies from the same size as a hindrance to share information, which makes the sharing with larger ones easier. The most important information sharing partners are located in several areas, mainly in Lisbon and in the north of the country.

The communication channels are both virtual and face-to-face, with a predominance of the first one. The most virtual channels used are e-mail, Skype and telephone. No inter-organizational information system was mentioned. One respondent reported a technology platform specifically designed to assist customers. Among the face-to-face channels, formal and informal meetings were mentioned. Respondents stressed the importance of face-to-face meetings during periods of contract negotiations and informal meetings as a way to get a closer relationship with the companies. Respondent D also stressed the fact that the most important information sharing partners of its company are fruit of previous personal relationship that developed into a business relationship.

Three kinds of information are shared among the companies and the most important information sharing partners:

- General market information in the case of relationship with companies from the same sector.
- Suggestions for improvements of products and services and specific features of the companies in the cases of relationships with clients.
- Information on the partner's technology in the case of joint projects.

Regarding the latter, when asked about the decision to keep the partnership rather than mastering the technology, interviewees B and D said that despite having some knowledge about the technology of the partners, the decision to maintain a partner is taken from a strategic point of view. One of them pointed out the fact that the partner's knowledge is much broader, which makes it more interesting to keep the partnership instead of investing resources to master the knowledge. The other one noted the lack of internal activities that justify the internalization of the partner's knowledge.

Finally, characteristics related to the interviewed companies, its partners and the relationship between them that influence the sharing of information between them were identified. The factors reported were partner's knowledge, trust, and openness to receive information, capability to establish relationships, absorptive capacity and an attitude of information protection.

The scope and domain of knowledge offered by the partner was relevant in all types of relationships. Trust emerged as an important factor in relations with customers and the partners in joint projects. In both cases trust is related to the expectation that the partner will not behave opportunistically, either passing the information on to competitors or not sharing the results equitably. For COTEC's coordination representative, trust in the partners is also important and can pass firstly through the intermediate role of the coordination, meaning that it's believed that the leadership of the network functions as a seal, a guarantee that companies belonging to the network are trustworthy. The trust in the leadership is a crucial aspect for the success of SME networks in Portugal, according to Moreira (2007).

Willingness to learn is related to what Hamel (1991 *apud* Simonin 2004) meant by the intention to learn, that is, the desire and will of an organization to learn from the partner. Simonin (2004), investigating international strategic alliances, reported that the intention of learning has a positive and significant effect on knowledge transfer, regardless of whether companies are competing with each other or not. According to Martinkenaite (2011), studies show the major importance of the receiver's motivation to acquire new knowledge than the donor to share knowledge.

Ability to establish relationships is related to what Ritter and Gemünden (2003) call skills to start inter-organizational relationships. According to the authors, the start of relationships demands efforts and investments such as the monitoring of the environment and the distribution of information about the company. The attitude of protecting information, understood as an inability or unwillingness to share (Simonin 2004), appeared as a factor that should be taken into consideration by companies and that inhibits sharing between firms from the same sector and size:

> I do not pass this knowledge willingly to a company, for example in the IT sector... It does not mean that in a partnership, it could be passed, if we assessed the risks very well... (Interviewee B)

Despite advances regarding internal organization, the absence of absorption capacity is still a problem. According to Cohen and Levinthal (1990, p. 128), absorptive capacity is the ability of an organization to "recognize the value of external information, assimilate it and apply it to commercial ends". In terms of identification of external information, only respondent B reported having a strategic interfaces management. Interfaces are understood as entities that can provide relevant knowledge to the company. In terms of assimilation, respondent B also identified that it is not always easy to understand the information shared by customers, considering the specific reality of each one, which requires from him "a significant effort of approximation". This fact stresses the importance of prior knowledge to facilitate the process of assimilation. In terms of storage, the most common practices cited were the registration of information related to business proposals, product documentation, events and conferences. Finally, incorporation seems to be, according to respondent D, the greater and more relevant internal deficiency:

> It has to do with the systematization of processes, methodologies, tools, repositories, to be sure that what we do is used, is incorporated in the chain and it is known to all ... innovation is not putting a team of development engineering to make a project, make the justification, develop, demonstrate...No. This has no use at all. The important is to take it and improve my offer (D).

Despite the limitations regarding absorptive capacity and the difficulties mentioned by some members to measure the benefits of information sharing with the most important partners, it was possible to identify the improvement and creation of new products as the main gains obtained by the interviewed companies both in the relationship with customers and in the cases of joint production.

5 Conclusions

This paper aimed to obtain a broader understanding of how and why information sharing occurs within an Innovation network. Under the collective sharing of information, it was possible to highlight the role of coordinating as a sender of relevant information to other members and also as a promoter of information sharing among them. It was also noteworthy a passive or even disinterested attitude from companies regarding the seeking of new relationships within the network. This passivity can be justified, firstly, by the difficulty of companies to identify mutual interests, which may indicate that either they do not know each other well enough or they are not sure about their learning goals within the network. The second reason may be the unwillingness to establish and maintain relationships with companies that belong to the network, which may indicate a lack of interest in learning or in giving information to each other. Considering that the network studied consists of companies classified as innovative, it is possible that the aim of gaining status and legitimacy for belonging to the network overlaps the search for learning. Regarding the main information sharing relationships, it was possible to notice that the partners differ in size, sector and type of relationship. It was also noteworthy that even being classified as innovative, there is still room for improvement in terms of absorptive capacity in the companies. Finally, this is a qualitative and exploratory study that, holding some limitations, requires a broader sample and the use of quantitative methodology in order to deepen the knowledge about the phenomenon.

References

Bond III, E.U., Houston, M.B., Tang, Y.: Establishing a high-technology knowledge transfer network: The practical and symbolic roles of identification. Industrial Marketing Management 37, 641–652 (2008)

Britto, J.: Elementos estruturais e conformação interna das redes de empresas: desdobramentos metodológicos, analíticos e empíricos (2001), http://www.race.nuca.ie.ufrj.br/sep/eventos:enc2002/m24-britto.doc (accessed September 20, 2002)

Capurro, R.: Epistemologia e Ciência da Informação. V Encontro Nacional de Pesquisa em Ciência da Informação. Belo Horizonte, Brasil (2003), http://www.capurro.de/enancib_p.html (accessed May 07, 2010)

Cohen, W.M., Levinthal, D.A.: Absorptive Capacity: A New Perspective on Learning and Innovation. Administrative Science Quarterly 35, 128–152 (1990)

Cotec Network Website, http://www.cotecportugal.pt/index.php?option=com_advfrontpage&lang=en,which

Creswell, J.W.: Research desing: qualitative and quantitative approaches. Sage publications, California (1994)

Dahl, M.S., Pedersen, C.R.: Knowledge flows through informal contacts in industrial clusters: myth or reality? Research Policy 33(10), 1673–1686 (2004)

Easterby-Smith, M., Lyles, M.A., Tsang, E.W.K.: Inter-Organizational Knowledge Transfer: Current Themes and Future Prospects. Journal of Management Studies 45(4) (2008)

Ernst, D., Kim, L.: Global production networks, knowledge diffusion and local capability formation: a conceptual framework. Paper Presented at the Nelson & Winter Conference in Aalborg, Denmark (January 26, 2012), http://www.druid.dk/conferences/nw/paper1/Ernst_and_Kim.pdf

Fiala, P.: Information sharing in supply chains. Omega 33 (2004), http://www.sciencedirect.com (accessed January 27, 2012)

Fritsch, M., Kauffeld-Monz, M.: The impact of network structure on knowledge transfer: an application of social network analysis in the context of regional innovation networks (2008), http://www.wiwi.uni-jena.de/uiw/publications/pub_since_2007.html (accessed April 20, 2010)

Hau, L.N., Evangelista, F.: Acquiring tacit and explicit marketing knowledge from foreign partners in IJVs. Journal of Business Research 60, 1152–1165 (2007)

Lawson, B., Petersen, K.J., Cousins, P.D., Handfield, R.B.: Knowledge Sharing in Interorganizational Product Development Teams: The Effect of Formal and Informal Socialization Mechanisms. Journal of Product Innovation Management 26, 156–172 (2009)

Madlberger, M.: What drives Firms to Engage in Interorganizational Information Sharing in Supply Chain Management? International Journal of e-Collaboration 5(2), 18–42 (2009)

Martinkenaite, I.: Antecedents and consequences of interorganizational knowledge Transfer: Emerging themes and openings for further research. Baltic Journal of Management 6(1), 53–70 (2011)

Mei, S., Nie, M.: Relationship between Knowledge Sharing, Knowledge Characteristics, Absorptive capacity and Innovation: an empirical study of Wuhan Optoelectronic Cluster. The Business Review 7(2) (summer, 2007)

Moberg, C.R., Cutler, B.D., Gross, A., Speh, T.W.: Identifying antecedents of information exchange within supply chains. International Journal of Physical Distribution & Logistics Management 32(9), 755–770 (2002)

Moreira, P.S.: Liderança e Cultura de Rede em Portugal: casos de sucesso. Livros Horizontes Ltda, Lisboa (2007)

Ritter, T., Gemunden, H.G.: Network competence: its impact on innovation success and its antecedents. Journal of Business Research 56, 745–755 (2003)

Simonin, B.L.: An empirical investigation of the process of Knowledge transfer in international strategic alliances. Journal of International Business Studies 35, 407–427 (2004)

Tushman, M.L., Scanlan, T.J.: Boundary Spanning Individuals: their role in information transfer and their antecedents. Academy of Management Journal 24(2), 289–305 (1981)

Wijk, R., Jansen, J.J.P., Lyles, M.: Inter- and Intra-Organizational Knowledge Transfer: A Meta-Analytic Review and Assessment of its Antecedents and Consequences. Journal of Management Studies 45(4) (2008)

Yin, R.K.: Case study research: design and method, 3rd edn. Sage Publications, London (2009)

Zhang, Q., Vonderembse, M.A., Lim, J.S.: Spanning flexibility: supply chain information dissemination drives strategy development and customer satisfaction. Supply Chain Management: An International Journal 11(5), 390–399 (2006)

Interdisciplinarity: Creativity in Collaborative Research Approaches to Enhance Knowledge Transfer

Janet Coulter

Research Institute for Art and Design (RIAD), University of Ulster, Belfast, UK

Abstract. This paper outlines how collaborative approaches were used to successfully generate research opportunities and enhance knowledge transfer in the development of new products. The researchers were drawn from across four disciplines within the University of Ulster and worked closely with an industrial partner to develop and test new approaches to the design and development of soft body armour. The paper highlights the drivers needed to support interdisciplinary collaboration and examines motivations for co-operation between industry and universities in the transfer, exchange and management of knowledge. A research project, highlighting two case studies demonstrates that design- and science-based partnerships can be successful in creating and transferring new knowledge.

Keywords: Interdisciplinarity, Collaboration, Knowledge Management, Knowledge Exchange, Innovation.

1 Introduction

The Company partner cited in this paper is a manufacturer of soft body armour. At the time of the research, the Company was engaged with The University of Ulster in a 27 month Knowledge Transfer Project (KTP) [1] to develop new products for new markets. In developing solutions for the Company, an in-house research and development (R&D) capacity was established. However, the Company had little design experience and some ideas suggested for R&D were outside the scope of the agreed objectives for the KTP. External funding was obtained by the author to research and explore ideas, which could potentially benefit the KTP. These projects were significant as they drew from diverse sources of knowledge. Expertise from fashion and textiles, sports and exercise, health and rehabilitation sciences and engineering all contributed to the research. This unique collaboration yielded interesting results and demonstrated how a cross-disciplinary approach to research can enhance knowledge transfer and contribute to economic growth.

The paper is contextualized with an examination of the reasons why academics choose to engage in collaboration and suggests the characteristics that typical cross-collaborators should possess. It outlines drivers for interdisciplinary alliances and motivations that steer industry to seek partnerships with universities

R.J. Howlett et al. (Eds.): *Innovation through Knowledge Transfer 2012*, SIST 18, pp. 169–178.
DOI: 10.1007/978-3-642-34219-6_19

and highlights the challenges that design-led and science-based partnerships need to overcome to reach successful outcomes.

2 Rationale for the Research

Strategic commitment is vital for innovation to be achieved (Rixon, 2003) and the Company was aware that R&D held the key. R&D feeds design with new technologies, materials and processes (Cooper and Press 2009) and provides strategic direction to inform potential research for new products. Cohen and Levinthal (1990) argue that companies with R&D capabilities are better equipped to utilise external information effectively. As the KTP project continued to generate new R&D, the Company realized that external knowledge would be required to research and explore the ideas. In seeking opportunities to create new knowledge without investment in new technology, the key challenges were to determine if: (i) existing technology within the University could be applied to the new design processes to provide novel solutions that could enhance competitive advantage and (ii) to establish how any perceived competitive advantage could be measured. The two case studies described in this paper demonstrate how, if a diverse knowledge skill set can be harnessed effectively, an open-innovation approach can produce successful results.

3 Case Studies

3.1 Case Study 1: 'SweatSmart'

New knowledge of technical textiles had been embedded in the Company as part of the KTP and the Company now sought to exploit this with the intention of creating a USP for its new product range. One issue deemed essential to explore was moisture management. Armour solutions have up to fifteen layers of Kevlar, plus laminates sealed inside a polyurethane cover, creating a thickness of approximately 25mm. These are then inserted into the back and front of a carrier vest, causing poor wicking properties. This combined with high external temperatures that military personnel may be subjected to, or stressful situations such as riot control, cause the wearer to perspire heavily. This can potentially lead to dehydration and disorientation, resulting in the wearer having reduced judgment and cognitive performance abilities. A study on dehydration in soldiers from tropical regions of India concluded that significant decrease in mental performance occurs at dehydration levels of 2% (Gopinathan, 1988). Lieberman (2007) provides a comprehensive review of the study and Cian et al (2000) cite increased fatigue and reaction time to decision-making and decreased short-term memory resulting from dehydration.

The investigative research, dubbed *'SweatSmart'* was initiated to determine whether innovative design, combined with performance fabrics could be engineered to improve moisture management and potentially reduce the associated physiological effects. An experiment was devised (Coulter et al, 2009) and implemented in a controlled environment in the University of Ulster, using an acclimatisation chamber at the Sport & Exercise Sciences Research Institute (SESRI). Researchers at SESRI introduced the scientific methodology of randomised controlled trials, which seek to completely remove extraneous variables without the researchers having to isolate them. The scientific advantage of randomised experiments is that it completely removes conscious or subconscious bias from the researcher and maximizes external validity [2].

Six prototype armoured vests were engineered, each using different design modifications and various combinations of technical textiles and spacer fabrics to assess which combination if any, would potentially increase airflow between the body and the armour, thus reducing dehydration. Prototypes were tested over a six-week period on sports studies students who exercised in the chamber for one hour at each session, working at 65% intensity in 70% relative humidity. These variables were determined by SESRI as the conditions most likely to mimic the heat dissipation felt by officers on routine, active duty. The prototypes were weighed pre- and post-exercise to measure moisture absorption by the jackets and the subjects were also weighed, pre- and post-exercise. Additional data was collected and measured including urine, temperature, thirst, heart rate and Borg scale data (Borg, 1982). Robust statistical measurements were applied to the data set and used to determine the best combination of design and technical fabrics to be used in the final prototype. The new knowledge was made available to the Company to allow it to embed it into its design strategy and enhance its products.

3.2 Case Study 2: Moulded Thermoplastic Composites for Female Body Contouring

Traditional construction methods of stitching darts into female body armour meant that with over 15 layers of Kevlar needed for each garment, the stitchers in the Company had to sew more than 30 darts into the finished vests, which was not only labour intensive, but additionally had crude results due to the stiff, laminated nature of the material, causing discomfort and reduced aesthetic appeal. Collaborative research between a fashion designer and an engineer set out to investigate whether multiple layers of Kevlar could be moulded three-dimensionally over a double curvature, in one operation to create a more ergonomically pleasing shape for female wearers.

Moulded, thermoplastic composites have superior impact and damage resistance properties, theoretically allowing lighter, protective garments to be manufactured to a nett-shape, thus providing a greater level of comfort for female users. In addition to their superior properties, it was estimated that thermoplastic composite vests could lead to more efficient production times than conventional pre-laminated, stitched products and could potentially make thermoplastic vests cheaper to produce than existing garments. Recyclability properties also had potential to save on waste management costs. Coulter and Archer (2009) explored the development of a process protocol for nett-shape, thermoplastic,

Kevlar-reinforced vests. The overall objectives were to design and manufacture a demonstrator component and to develop a manufacturing route for an optimised armour vest for females. The engineer determined variables in thermoplastics and temperature and the fashion designer determined variables in fabric structure and pile, allowing initial tests to be carried out to establish which combinations would positively affect the success of the outcome. The engineering input provided vital knowledge and expertise in thermoplastics, but lacked sensitivity to the anthropometrics of female anatomy. Once the concept of 3D Kevlar moulding was proven to work using a basic, double-curvature, domed template, more accurate data on female anatomy was obtained by the fashion designer using a body scanner at the Health and Rehabilitation Sciences Research Institute (HRSRI) at the University of Ulster. (HRSRI) followed ethical procedures and live subjects were used to create images, from state-of-the-art, 3D imaging technology. Applying knowledge in anthropometrics, the fashion designer extrapolated the data created by HRSRI and developed a blended dataset so that the final torso measurements were more generic, thus ensuring that the finished fit of the moulded Kevlar would span a greater size range of females.

The new knowledge from the dataset was translated by the engineer via CAD CAM into a machined torso in aluminum, which was then used to successfully mould multi-layers of Kevlar on to it in one operation, using thermoplastic techniques. The unique new knowledge was presented to the Company, allowing them to make informed decisions about the innovative development of their products.

4 Managing Knowledge in University-Led Collaboration

The Company partner was keen to build on the new university knowledge to develop strategies to help to survive the ongoing recession. Evidence suggests that organisations surviving the recession are doing so by being innovative in their business models and their approach to design. In addition to strong customer focus, knowledge management provides the key to successful innovation and Egbu et al (1999) contend that collaboration is vital for knowledge management. There are strong synergies between theories of strategic management and knowledge management and understanding their relationship provides a useful structure for understanding the importance of collaborative research partnerships between universities and industry.

Knowledge exchange between companies and universities can take many forms, ranging from collaborative R&D projects to informal agreements undertaken on an ad hoc basis (Cassiman & Veugelers, 2005). The Chesbrough (2003) open innovation model, suggests that companies should use external ideas as well as internal ones to further innovation. Bruce and Morris (1998) agree, contending that companies seek partnerships when their own internal design capabilities become complacent and stale and lack the vision and innovation that external design expertise can provide. Knudsen (2007) concurs with the author, arguing that external collaborators can provide specific, technical knowledge for bespoke projects as in the two outlined case studies in this paper. Harnessing external collaboration can provide companies with a suite of diverse and complementary resources necessary to turn innovative ideas into successful, commercial outputs (Hagedoorn, 1993) and this explains why companies seek to develop sources of knowledge with external partners such as universities, as an effective means of delivering core benefits (Baker & Sinkula, 1999). The specialised support that academic knowledge can provide can be accessed by emerging technologies and state-of-the-art equipment to further their own internal R&D make university partnerships an attractive option to industry (Tidd & Trewhella, 2002). Universities are seen as low-risk sources of innovation that are especially useful for new product development and their research institutes can generate a wealth of knowledge (Brettel and Cleven, 2011). Tijssen (2002) and Narin et al (1997) concur and evidence shows that companies that do not acquire technological knowledge from universities may fall behind with innovation and be less likely to make technological breakthroughs that lead to viable commercial products (Spencer 2003).

The new knowledge that the Company needed could not be achieved with one type of expertise alone, hence the interdisciplinary approach and Jobber (1991) argues that combining research procedures is more useful than one single procedure, providing a more comprehensive analysis of the problem studied. Rhoten and Pfirman (2007) define interdisciplinarity as "the integration or synthesis of two or more disparate disciplines, bodies of knowledge, or modes of thinking to produce a meaning, explanation, or product that is more extensive and powerful than its constituent parts."

5 Drivers for Collaboration between Universities and Industry

Understanding the motivations and processes leading academics to engage in collaborative research and the resultant technological developments provide vital strategic insights for companies, allowing them to develop innovation strategies. Mueller (2007) concurs, contending that entrepreneurship and university–industry relations are vehicles for knowledge flow and thus spur economic growth.

Perkman and Walsh, (2009) devised a classification framework for types of university-industry collaborations which suggested that those focused toward 'ideas testing' are typically low-cost projects initiated by either academics or companies to investigate potentially interesting and commercial ideas, as outlined in the two case studies in this paper. Narin et al (1997) also cites cost effectiveness as a key factor for university-industry engagement. Kabins (2011) suggests that a motivating factor for academics engaging in university-driven collaborations which are aimed at testing ideas, may be the close involvement of public sectors end users and the inspiration for academics becoming involved is the quality of data collected when investigating 'live' problems and genuine needs, whereas companies tend to be more passively engaged and interested solely in the end product and commercial output. Having the MoD as an end user of the Company's products was a strong driver for academic involvement in this project.

It is important to balance some of the reviews that may over-simplify the motivations and execution of interdisciplinary collaboration. There are many drawbacks which need to be given due consideration. Often there are more costs, greater co-ordination requirements and project management challenges. Katz and Martin, (1997) and Nooteboom (2000) caution that costs can escalate due to the cognitive distance between partners. It was established at the outset of this research that costs and logistical factors would not outweigh the benefits of collaboration, however the main challenges would be in understanding the working methodologies and 'languages' of practice from the different disciplines involved.

6 Characteristics of Academic Collaborators

There is limited research data available on the characteristics of individuals who choose to engage in interdisciplinary collaboration. The motivation for some researchers is simply, to learn from distant disciplines of science or because they have identified a specific area of research that requires the input from disparate disciplines, as was the case with this author. They may also be driven by the possibility of future benefits in terms of publications, recognition and further funding opportunities (Melin, 2000). Van RijnSouever and Hessel (2011) and Carayol (2004) define a set of characteristics of researchers engaging in interdisciplinary collaboration. However Rafols and Meyer, (2007) cite consolidating knowledge where problems to be solved are complex and increasing access to funding opportunities as equally valid drivers. Enhancing personal credibility and desire to earn peer recognition are also contributing factors (Whitley 2000).

7 Dynamics and Challenges of Interdisciplinary Collaboration

There are limited examples of successful collaborations between design-based subjects such as fashion and more science-based subjects, such as engineering. One reason for this is perhaps how research is framed and how outcomes are measured and disseminated differently in each discipline. Scientific methodologies rely on defining a set of principles and procedures that are used by researchers to develop questions, collect quantitative data and reach conclusions. However, design-based research relies on strategies of qualitative inquiry and outcomes are often measured in a more ambiguous way, such as user perception and emotional tactility and these metrics do not always sit comfortably within the science domain. These different 'languages' of research can lack commonality and understanding, which can make knowledge difficult to transfer and may inhibit collaborative success. Design- and arts-based research does not always fit in with existing typologies of scientific research and new typologies of qualitative research need to be recognised by scientists to bridge the gap and establish a greater respect for art and design based research (McNiff, 1988). Similarly, designers need to make a culture shift and embrace scientific methodologies. The author contends that it is likely that designers will need to make the greater shift and will have to accept new technology and develop a better understanding of scientific methodologies. Openness to a range of diverse research methodologies, willingness to engage, understand and demonstrate mutual respect for all contributors are required to deliver successful outcomes. This is not always an easy task for scientific, left-brain thinkers who prefer well-structured, well-defined tasks, to open-ended ones. Design processes need to be adaptable and allow for ambiguity and whilst designers' right-brain thinking skills are adept to these ad hoc processes of working, the uncertainty of outcomes can be challenging for engineers and scientists to accept. Fraser (2010) stresses the importance of collaborative researchers being able to embrace both the friction and the fusion that come with intense collaboration.

There are pros and cons regarding case studies and pure statistical methodologies and whilst physical scientists tend to avoid a case study approach, anthropologists consider them an essential tool. A case study may introduce unexpected results, leading to new avenues of research and it is also recognised that in the dissemination of results, case studies provide more interesting themes than purely statistical surveys. While statistics may be interesting to scientists or other academics, well-positioned case studies can give a stronger impact to a wider audience.

8 Conclusion

Within the context of the case studies outlined in this paper, it is concluded that the combination of design- and science-based research succeeded in creating new knowledge, which was successfully transferred to industry. Equally important outcomes to the project were the knowledge exchange and innovative approaches to research and design acquired by the collaborators. The designer clearly had much to learn from scientific approaches and methodologies, including randomized trial experiments and scientific measurement of outcomes. Knowledge was gained of technologies within engineering and rehabilitation sciences and new codes of practice, such as ethical approval were noted. The scientists benefitted greatly from design-led approaches and working creatively, in addition to gaining knowledge of high-performing technical textiles, anthropometrics and most significantly commercial application.

It is anticipated that as more designers embrace technology then more collaboration and innovative knowledge transfer will be possible. Cohen et al. (2002) note that generating scientific knowledge together with the associated benefits that universities can offer is a key foundation of industrial innovation and Schmickl and Kieser (2008) note that interdisciplinary science has a positive effect on knowledge production and innovation.

It is worth noting that, had the project spanned a longer period of time and had a larger budget, there may well have been more challenges for leadership and management within the project.

This interdisciplinary project was initiated and led by a fashion designer. Designers are imaginative experimenters who often take on a 'cross-pollinator' role (Kelley, 2004) that allows them to work creatively across many disciplines. The key for a designer always lies in creativity and as a designer from a non-scientific background the author offers a metaphoric 'creative equation' as a tool to begin build a common language for scientists and designers. The premise being that the greater the multiplier of creativity, in science-based collaboration, then the greater the likelihood of innovative new knowledge created:

(SK+ACM) x C = INK

where **SK = Scientific Knowledge**
ACM = Appropriate collaborative methodologies
C= Creativity
INK = Innovative new knowledge

Source: Coulter 2011

Acknowledgements.
Dr. Edward Archer, ECRE Engineering Composites , Research Centre, University of Ulster
Dr. Gareth Davison, Sport & Exercise Sciences Research Institute, University of Ulster
Dr. Conor McClean, Sport & Exercise Sciences Research Institute, University of Ulster
Prof. John Winder, Health and Rehabilitation Sciences Research Institute, University of Ulster
Mrs Orlagh Carmody, Head of Design, Global Armour Ltd
Mr Gary Hayes, Sales and Marketing Director, Global Arnour Ltd
Mr Mark Griffiths, Sales Manager, Woven Industrial Solutions, Heathcoat Fabrics, Devon.

References

Borg, G.A.: Psychophysical bases of perceived exertion. Medicine and Science in Sports and Exercise 14, 377–381 (1982)

Baker, W.E., Sinkula, J.M.: Learning orientation, market orientation, and innovation: Integrating and extending models of organizational performance. Journal of Market Focused Management 4, 295–308 (1999)

Morris, B.: A comparative study of design professionals. In: Bruce, M., Jevnaker, B.H. (eds.) Management of Design Alliances: Sustaining Competitive Advantage. John Wiley, Chichester (1998)

Cassiman, B., Veugelers, R.: R&D Cooperation between Firms and Universities: Some empirical evidence from Belgium (pdf). The International Journal of Industrial Organization 23(5-6), 355–379 (2005)

Carayol, N., Thuc Uyen Nguyen, T.: Why do academic scientists engage in interdisciplinary research. Research Evaluation 14(1), 70–79 (2005)

Cian, C., Koulmann, N., Barraud, P., Raphel, C., Jimenez, C., Melin, B.: Influence of variations in body hydration on cognitive function: effect of hyperhydration, heat stress, and exercise-induced dehydration. J. Psychophysiology 14, 29–36 (2000)

Chesbrough, H.W.: The Era of Open Innovation. MIT Sloan Management Review 44, 35–41 (2003)

Cohen, W., Levinthal, D., Damiel, A.: Absorptive Capacity: A New Perspective on Learning and Innovation. ASQ 35, 128–152 (1990), `http://faculty.babson.edu/krollag/org_site/org.../cohen_abcap.htm` (accessed January 10, 2012)

Cooper, R., Press, M.: The Design Agenda; A guide to Successful Design Management. John Wiley & Sons, UK (2000)

Coulter, J., Davison, G., McCLean, C., Carmody, O.: "What if?' Research and Innovation by Design. In: Proceedings from the Senior Staff Conference, University of Ulster (2009)

Coulter, J., Archer, E.: "What if?' Research and Innovation by Design. In: Proceedings from the Senior Staff Conference. University of Ulster (2009)

Daanen, H.A.M., Reffeltrath, P.A., Koerhuis, C.L.: Ergonomics of Protective Clothing. In: Jayaraman, S., et al. (eds.) The Netherlands in Intelligent Textiles for Personal Protection and Safety. IOS Press, Amsterdam (2000)

Dalpe, R., DeBresson, C., Xiaoping, H.: The public sector as first user of innovations. Research Policy 21, 251–263 (1992)

Egbu, C.O., Sturges, J., Bates, M.: Learning from Knowledge management and trans-organisational innovations in diverse project management environments. In: Hughes, W. (ed.) 15th Annual ARCOM Conference, Association of Researchers in Construction Management, September 15-17, vol. 1, pp. 95–103. Liverpool John Moores University (1999)

Fraser, H.: Designing business. New models for success. In: Lockwood, T. (ed.) Design Thinking: Integrating Innovation Customer Experience and Brand Value. Allworth Press, New York (2010)

Gopinathan, P.M., Pichan, G., Sharma, V.M.: Role of dehydration in heat stress-induced variations in mental performance. Arch. Environ. Health 43, 15–17 (1988)

Hagedoorn, J., Duysters, G.: External Sources of Innovative Capabilities: The Preferences for Strategic Alliances or Mergers and Acquisitions. Journal of Management Studies 39, 167–188 (2002)

Huxman, Vangen: Managing to collaborate. Theory and practice of collaborative advantage. Routhledge, London (2005)

Jobber, D.: Choosing a survey method in management research. In: Smith, N.C., Dainty, P. (eds.) The Management Research Handbooks. Routledge, London (1991)

Kabins, S.: Evaluating outcomes of different types of university-industry collaboration in computer science. In: Conference Proceedings Academy of Management (AOM) Conference (2011)

Katz, J.S., Martin, B.R.: What is research collaboration? Research Policy 26, 1–18 (1997)

Kelley, T., Littman, J.: The Ten Faces of Innovation; Strategies for heightening Creativity. Profile Books (2004)

Knudsen, M.P.: The Relative Importance of Interfirm Relationships and Knowledge Transfer for New Product Development Success. Journal of Product Innovation Management 24, 117–138 (2007)

Lieberman, H.R.: Hydration and Cognition: A Critical Review and Recommendations for Future Research. J. Am. Coll. Nutr. 26(suppl. 5), 555S–561S (2007)

McNiff, S.: Arts Based Research Arts Based research. Jessica Kingsley Ltd., London (1988)

Melin, G.: Pragmatism and self-organization – research collaboration on the individual level. Research Policy 29(1), 31–40 (2000)

Mellis, W.: Partnering with Customers. In: Jolly, A. (ed.) Innovation; Harnessing Creativity for Business Growth. Kogan Page, London (2003)

Mueller, P.: Exploring the knowledge filter: How entrepreneurship and university–industry relationships drive economic growth. Research Policy 35(10), 1499–1508 (2007)

Narin, F., Hamilton, K.S., Olivastro, D.: The Increasing Linkage between US Technology and Public Science. Research Policy 26, 317–330 (1997)

Nooteboom, B.: Learning by interaction: absorptive capacity, cognitive distance and governance. Journal of Management and Governance 4, 69–92 (2000)

Perkmann, M., Walsh, K.: University-industry relationships and open innovation: Towards a research agenda. International Journal of Management Reviews 9, 259–280 (2007) (review article)

Perkmann, M., Walsh, K.: How firms source knowledge from universities: partnerships versus contracting. In: Bessant, J., Venables, T. (eds.) Creating Wealth from Knowledge, Edward Elgar, Cheltenham (2007) (forthcoming)

Press, M., Cooper, R.: The Design Experience, The Role of Design and Designers in the Twenty First Century. Ashgate Publishing Limited, Hampshire (2005)

Rafols, I., Meyer, M.: How cross-disciplinary is bionanotechnology? Explorations in the specialty of molecular motors. Scientometrics 70(3), 633–650 (2007)

Rhoten, D., Pfirman, S.: Women in interdisciplinary science: exploring preferences and consequences. Research Policy 36(1), 56–75 (2007)

Rixon, D.: Strategic Commitment- The Foundation Stone for Innovation. In: Jolly, A. (ed.) Innovation; Harnessing Creativity for Business Growth. Kogan Page, London (2003)

Schmickl, C., Kieser, A.: How much do specialists have to learn from each other when they jointly develop radical product innovations? Research Policy 37(3), 473–491 (2008)

Spencer, J.W.: How Relevant is University-Based Scientific Research to Private High-Technology Firms? A United States–Japan Comparison. Academy of Management Journal 44, 432–440 (2001)

Tidd, J., Trewhella, M.J.: Organizational and Technological Antecedents for Knowledge Acquisition and Learning. R&D Management 27, 359–375 (2002)

van Rijnsoever, F.J., Hessels, L.K.: Factors associated with disciplinary and interdisciplinary research collaboration. Research Policy 40, 463–472 (2011)

Whitley, R.: The Intellectual and Social Organization of the Sciences, 2nd edn. Oxford University Press, Oxford (2000)

[1] http://www.ktponline.org.uk/strategy/

[2] http://www.experiment-resources.com/randomized-controlled-trials.html#ixzz1moF2sRxd

Requirements of Knowledge Commercialization in Universities and Academic Entrepreneurship

Samira Nadirkhanlou[1], Ali Asghar Pourezzat[2], Arian Gholipour[2], and Mona Zehtabi[3]

[1] University of Tehran, Tehran, Iran
[2] Department of Public Administration, Faculty of Management, University of Tehran, Tehran, Iran
[3] Management Department, University of Tehran
{nadirkhanlou,pourezzat,agholipor,zehtabi}@ut.ac.ir

Abstract. Knowledge as the main competitive advantage in the world economy has very important role in countries development. So it seems necessary for the universities, as the main institution of knowledge generation, to participate in national and regional economic development. compiling a proper pattern of academic entrepreneurship and commercial knowledge transfer becomes more evident if the need for income earning and knowledge commercialization for survival of the universities and existence of barriers to integral exploitation of universities' intellectual properties are considered. In this paper we tried to provide a model for the commercialization of knowledge in universities. Regarding the commercial knowledge transfer methods of five top universities of the world according to the ranking of The Times Higher Education-QS 2007 (Harvard, Yale, Oxford, Cambridge and Imperial College London); this research is devoted to compiling a benchmarked pattern. The factors extracted from this pattern are then ranked according to experts' viewpoints using analytic hierarchy process. According to the findings of this study "Adopting incentive policies in royalty sharing for faculties" is most important from the perspective of academic entrepreneurship and knowledge commercialization experts, and "Networking", "financial support", "creating the necessary structures" and "Faculty freedom" are placed in the next priorities, respectively.

Keywords: academic entrepreneurship, knowledge commercialization, intellectual property, spin offs, license.

1 Introduction

Many changes have been made in today's universities in comparison to the universities in the past decades, and accordingly expectations from the universities have also been changed. The first academic revolution, taking place in the late 19th century, made research a university function in addition to the traditional task of teaching (Etzkowitz, 2003). In the late 1970s, a reconceptualization of the role of public research systems commenced in the U.S. following a growing concern

R.J. Howlett et al. (Eds.): *Innovation through Knowledge Transfer 2012*, SIST 18, pp. 179–194.
DOI: 10.1007/978-3-642-34219-6_20

about the apparent deterioration of the national comparative advantage in high technology (Baldini et al., 2006) and, increasing criticism of universities for being more adept at developing new technologies than moving them into private sector applications (Siegel et al., 2003). This led to the second academic revolution in which contribution to economic and social development of communities was added to universities' functions and missions. These revolutions can be attributed to some factors; nowadays, science and knowledge play specific role in global competition, it has emerged as an alternative engine of economic growth to the classic triumvirate of land, labor and capital- the traditional sources of wealth (Etzkowitz, 2003). In fact, the emergence of new economy, which occurred during the 1990s, is altering the nature of relationships among science, technology, innovation and economic performance.

Emerging expectations about the direct contributions of academic institutions to economic growth, permission of universities for patent inventions, and creation of new organizational units specifically dedicated to technology transfer activities were among the effects of reforms in commercialization trend (Baldini et al., 2006).

However, this process does not always happen easily and successfully (Decter et al., 2007). As commercialization activities may affect both teaching and research activities of the universities, there is a potential for conflict and resistance (Rasmussen et al., 2006), thus in many cases, this process is not adequately effective and efficient.

Therefore, in this paper it was attempted to primarily examine methods of five top universities of the world based on The Times Higher Education-QS 2007 ranking (i.e. Harvard, Yale, Oxford, Cambridge and Imperial College London). This research is devoted to compile a benchmarked pattern, and the factors extracted from this pattern are ranked according to experts' viewpoints.

2 Entrepreneurial University

Some researchers refer to all commercialization activities out of usual research and educational functions in academic entrepreneurship definition (Klofsten and Jones-Evans, 2000); while others, emphasize more on university's emerging activities, especially spin off creation; for example, it can be referred to definition of Etzkowitz (2003) which supposes entrepreneurial university as a natural incubator with both commercial and intellectual supportive structures (or as combination of them), as faculties and students can set up new firms relying on these structures. Wright et al. (2007) consider academic entrepreneurship as a commercialization development beyond the traditional focus on licensing intellectual properties. They refer to it as containing a move to establish spin off companies which are the result of science and technology at universities. Also in some studies, academic entrepreneurship has been defined as an area of entrepreneurship which is looking for understanding and describing new ventures and productions due to university's intellectual property (Llano, 2006).

To capture transformations occuring among institutional spheres of community, two frameworks have been proposed: the National Systems of Innovation (NSI) and triple helix models. The NSI focuses on existing firms (as the innovation engine) and other organizations (as a support structure). While the focus of the triple helix is on interaction among university, industry and government as well as the creation of hybrid organizations, such as the incubator, to support the start-up process (Etzkowitz et al., 2005). It states that the university can play an enhanced role in innovation in increasingly knowledge-based societies. The underlying model is analytically different from the national systems of innovation (NSI) approach which considers the firm as having the leading role in innovation. Triple helix model focuses on the network overlay of communications and expectations that reshape the institutional arrangements among universities, industries, and governmental agencies (Etzkowitz and Leydesdorff, 2000).

3 Knowledge Commercialization

Several factors have caused universities to change their views on the instrumental aspects of research and make attempts to collaborate more actively with private companies. One is the search for additional sources of research funding (OECD, 2000).

In recent decades, two "waves" of commercialization have been identified. The first one which mainly commenced since the beginning of the 1980s can be recognized by the establishment of "traditional" science parks, most often aimed at attracting advanced companies. The second wave which was accelerated around the last half of the 1990s focuses on creation of spin-offs and licensing university intellectual properties and increased involvement by students in commercialization activities (Rasmussen et al., 2006).

4 Commercial Knowledge Transfer Requirements

Requirements of commercialization of academic research results can be divided into three categories; namely, cultural, structural, and political.

For development of university product commercialization, the relevant culture should be created and encouraged. Investigations in this context emphasize on infrastructural reforms and institutional innovations for creating and promoting an entrepreneurial and supportive culture within the academic institution (O'Shea et al., 2005; Henrekson and Rosenberg, 2001).

Traditional boundaries between university and industry are disregarded (Etzkowitz, 1998) and the increased emphasis on knowledge and technology transfer across university–industry institutional boundaries has led to the creation and implementation of a variety of transfer-oriented mechanisms. These include industrial liaison or technology transfer offices, academic spin offs and joint ventures (whereby universities start acting as a shareholder), science parks and business incubators (Looy et al., 2004). These mechanisms include development of industrial liaison offices, technology transfer offices, university spin offs, and joint ventures, science and technology parks, and incubators for providing supportive services to start ups (Grimaldi and Grandi, 2005) or joint ventures in which universities start acting as a shareholder (Tijssen, 2006). Indeed, creation of technology transfer offices in many universities has led to increase in university-industry relationship (Debackere and Veugelers 2005).

In 1980, US Congress attempted to remove potential obstacles of industry-technology transfer through legislation, which became known as the Bayh–Dole Act (Siegel et al.,2003). According to this act, the ownership of publicly funded research is shifted to the researcher organization, and this has created stronger incentives for universities to look for commercial applications of their intellectual properties (Debackere and Veugelers 2005). The impact of this act is so great that it is mentioned in about all academic entrepreneurship and commercialization studies and is considered as the most influential factor in development of academic entrepreneurship in the US. Through creating economic incentives for universities, this act allows them to experiment various ways and find the best way of economic development. It has no compulsion or even suggestion about appropriate response to commercial opportunities. While in most European Union countries, the governments directly dictate policies to universities and research institutes in order to promoteand facilitate the transfer of intellectual properties (Goldfarb and Henrekson, 2003).However, given the main role of knowledge-based entrepreneurship in the process of economic growth, policy-makers can encourage the transformation of generally available new knowledge into viable new products and technology by putting greater emphasis on entrepreneurship policy (Audretsch et al., 2008).

5 Methods of Knowledge and Technology Transfer

Regardless of being commercial or noncommercial, university knowledge and technology can be transferred to public and private sector organizations in various ways. Bozeman (2000) introduces eight different communication channels for technology transfer process: Literature, Patent, License, Absorption, Informal, Personnel Exchange, On-site Demonstration, and Spin-off. Cooperation with industry for development of products and services, provision of consulting services, meeting with industry staff, creating university technology transfer based economic firms, joint research with industry and so on are among knowledge transfer methods which are mentioned in various studies (Landry et al., 2007).

In general, commercial activities in universities are mainly performed in two ways: a. licensing intellectual properties and provision of consulting services for public and private sector (which is old and traditional form of these activities and have a relatively long age). b. creating university technology based new firms (which is a new and emerging form of knowledge, specially advanced technology commercialization and has recently attracted much attention) (Di Gregorio and Shane, 2003).

Universities can, according to their rules, make policies and strategies to create equity in the spin offs in exchange for intellectual property fees (Lockett and Wright, 2005). Spin off creation. as opposed to licensing, is an important entrepreneurial method for knowledge commercialization (Debackere and Veugelers 2005).

6 Research Method

Regarding the commercial knowledge transfer methods of five top universities of the world according to the ranking of The Times Higher Education-QS 2007 (Harvard, Yale, Oxford, Cambridge and Imperial College London), this research is devoted to compiling a benchmarked pattern. Factors of this pattern extracted through comparative study. These factors are then ranked according to experts' viewpoints.

6.1 Sample

Five top universities of the world were selected based on Times Higher Education Institution 2007 rankings (The Times Higher Education, 2007), i.e. Harvard University, Cambridge, Oxford, Yale, and Imperial College London, in order to make comparison of similarities and differences with respect to methods, policies and processes of commercial knowledge transfer.

Based on documents available from these universities and their content analysis, the similarities and differences were reviewed through comparative study method and a model was developed as "benchmarked pattern" and the main requirements of knowledge commercialization in universities were identified.

6.2 Respondent Characteristics

Questionnaires of this study were completed by experts. Academic entrepreneurship and knowledge commercialization experts and specialists were selected from among officials of Entrepreneurship Center, entrepreneurship faculties and the Deans or Vice-deans of research section of faculties at University of Tehran.

6.3 Instrument

In this study, a questionnaire was used. This questionnaire was distributed for a survey of experts and specialists of University of Tehran to get their viewpoints about features of desirable pattern and prioritize the extracted requirements of knowledge commercialization. Questions of this questionnaire were designed with the aim of paired comparison of factors introduced in the model to allow respondents to measure the priorities of factors in relation to each other. Results of this questionnaire were analyzed through Analytic Hierarchy Process. First, decision criteria were prioritize with respect to the aim, and then all alternatives compared in pairs with respect to the each criteria (see section 7.6).

6.4 Validity and Reliability

The questionnaire was reviewed and confirmed by several professors. If the inconsistency rate of a questionnaire was higher than the acceptable limit (0.1), the questionnaire was completed again by the respondents.

7 Comparative Study

7.1 Comparison of the Characteristics of Universities Technology Transfer Offices

Comparison of the mission of the selected universities' technology transfer offices indicates that *the public interest* is preferred in some of them, and income earning and economic development, defined in line with public interest, are emphasized next.

Table 1 Services Provided by Technology Transfer Offices[1]

University	Services
Harvard	– licensing intellectual property resulting from research – assisting faculty to obtain patent protection for their technology – developing new ventures in related to Harvard's technology – providing advice to the Provost on matters pertaining to university policy on intellectual property
Oxford	– Services for helping researchers for: – Identifying research output of potential commercial value – Evaluating their commercial potential – Protecting research output with IPR – Marketing invention – Deal-making
Yale	– evaluating the commercial potential of inventions and discoveries – pursuing patents appropriately – patenting commercial partners to license and develop Yale technology
Cambridge	– Technology Transfer Services – Consultancy Services – Seed Funds and New Venture Services
Imperial College London	– Technology sourcing – Intellectual property management – Commercial assessment of intellectual property – Market analysis – License negotiation – Incubation services and space – Investment

[1] See:
- http://www.provost.harvard.edu/intellectual_property_and_licensing/techdev. php, 2008-10-09.
- "Intellectual Property, Patents AndLicences" (2007) Isis Innovation, university of oxford, UK. England, Oxford.
- "Yale Policy on Intellectual Property, Patents, and Licensing Agreements" (2002) Yale University, US.Connecticut. New Haven.
- http://www.enterprise.cam.ac.uk /aboutus.php, 2008-23-08.
- http://www.imperialinnovations.co.uk/?q=about/about-tertiary, 2008-13-08.

In Table 1, the main services provided by universities technology transfer offices have been compared.

7.2 Comparing the Ways of Managing the Intellectual Property

According to the principles of intellectual property management compared in Table 2, two American universities, Yale and Harvard, focus on principles that serve the interests of various stakeholders in exploitation of the intellectual property.

In these universities, commercialization is not regarded as an independent activity in order to manage intellectual properties, rather it is noted that commercialization or public dissemination paths will be selected and pursued based on public interests. Therefore, commercialization is recommended and supported while providing public interest.

Moreover, principles governing University of Cambridge's intellectual property management, managing conflict of interests in their exploitation, and accordance with academic missions have been mentioned.

Each university has mentioned terms in its intellectual property management policy for specifying ownership of intellectual property which are summarized in Table 3.

What is observed about determining ownership of university intellectual property is that, in general, using financial resources and equipment of university and production of intellectual property is the most principal criteria of its entitlement to university.

7.3 Comparing Issues Related to New Firms

In all universities, except Harvard, the business plan to create new companies based on university technology is provided by the technology transfer office. However, all of the five universities help researchers in identifying the management team and investors by creating networks among various parties involved in establishing new companies (such as venture capitalists, industrialists, entrepreneurs, and academic researchers).

Technology transfer offices at Harvard, Oxford and Cambridge provide substantial consultation services for interpretation and implementation of the rules and regulations for exploitation of intellectual property required by the researchers.

Table 2 Comparing Principles Governing Intellectual Property Management[1]

University	Principles governing intellectual property management
Harvard	Generic principles governing intellectual property management: – Considering public interest as well as inventors and authors interests; – Protecting the traditional rights of scholars with respect to the products of their intellectual endeavors; – Recognizing outside contractual commitments with respect to intellectual properties, where financial or other support for development of intellectual property has been provided by outside parties. Principles Governing Commercial Activities: – Primacy of the academic mission; – Freedom of inquiry; – Openness of inquiry; – Educational welfare of students; – Public trust.
Oxford	--------------------------------
Yale	– Strong patent protection in order to value creation to both university and society; – Focus on those opportunities with the highest probable benefit to both university and society; – Vigorously pursuing intellectual properties with high potential to improve the health or prosperity of the global community; – Value adding at every stage of the development process of intellectual property from discovery to market.
Cambridge	– Focusing on cases with the strongest potential to make a significant positive commercial impact; – Taking the course which supports commercialization of the technology; – Working effectively with inventors to support their aspirations, manage conflicts and encourage synergy with the mission of the University; – Finding the best partner to move the idea forward; – Negotiating the fair and reasonable terms which reflect the contribution of the assets and expertise being transferred; – Negotiating and making the greatest number of the best possible deals; – Looking after the deals once they are made to encourage commercialization and optimize returns
Imperial College London	--------------------------------

[1] See:
- "Statement of Policy in Regard to Intellectual Property", 2008, Harvard University, US. MA. Cambridge.
- http://www.yale.edu/ocr/about/index.html, 2008-12-08.
- http://www.enterprise.cam.ac.uk/aboutus.php?sub=131, 2008-23-08.

Table 3 Comparing Terms of University Entitlement to Intellectual Property[1]

University	Terms of university entitlement to intellectual property
Harvard	It is conducted by individuals covered by patent policy and has the following condition: – Under or subject to an agreement between Harvard and a third party; or – Be through direct or indirect financial support from Harvard, including support or funding from any outside source awarded to or administered by Harvard; or – With use (other than incidental use) of space, facilities, materials or other resources provided by or through the University.
Oxford	It depends upon the source of funding. Isis Innovation Limited ('Isis') shall be entitled to and responsible for the exploitation of research funded with research council grants.
Yale	– In the case of faculty and staff ordinarily any invention made by a Yale faculty member is owned by the University, unless: ▪ the invention is unrelated to the activities for which the individual is employed, and ▪ The invention was not made or conceived under circumstances involving University facilities or personnel. – In the case of students: ordinarily any invention is owned by the University
Cambridge	University is entitled, when intellectual property rights arise from the results of activities in the course of employment by the University.
Imperial College London	In the case of faculty and staff: inventions and other forms of IP generated by an employee belong to his/her employer if they are made in the course of the employee's normal duties. In the case of students: Where students generate IP in the course of their study or research they will own that IP in their own right unless one of the following applies: (i) they have a sponsored studentship under which the sponsor has a claim on arising IP; (ii) They participate in a research programme where arising IP is committed to the sponsor of the research; or (iii) They generate IP which builds upon existing IP generated by a member of academic staff, or is jointly invented with a member of academic staff. In the case of honorary staff: They are required to assign the rights to any IP they create in the course of their honorary activities to the University.

[1] See:
- "Statement of Policy in Regard to Intellectual Property", 2008, Harvard University, US. MA. Cambridge.
- Yale University Patent Policy" (1998) Yale University, US.Connecticut. New Haven.
- "Intellectual Property Rights" (2005) University of Cambridge, UK. England. Cambridge.
- "Intellectual Property Rights" (2005) Imperial College London, UK. London.

7.4 Comparing Royalty Income Sharing Policies

Types of stakeholders of royalty income in universities and how much or how their share of distribution are specified in Table 4.

Table 4 Comparing Stakeholders of Royalty Income in Universities[1]

University / stakeholders	Harvard	Oxford	Yale	Cambridge	Imperial College London
Inventor(s)	(personal share) 35% (Research share)12.75%	Diverse (decreasing)	Diverse (decreasing)	Diverse (decreasing)	Diverse (decreasing)
Technology transfer office	-	-	10%	Diverse (increasing)*	Diverse
Inventor's Department/Center	12.75%	Diverse (increasing)	-	7.5% (Above £50,000)	-
Inventor's school	17%	-	-	-	Diverse
University	9.75%	Diverse (increasing)	Diverse (increasing)	-	12.5% (Above £500,000)
University's investment funds	-	-	-	7.5% (Above £50,000)**	-
President	12.75%	-	-	-	-

[1] See:

- "Statement of Policy in Regard to Intellectual Property", 2010, Harvard University, US. MA. Cambridge.
- "Regulations for the Administration of the University's Intellectual Property Policy", 2011, Isis Innovation, university of oxford, UK. England, Oxford.
- "Yale University Patent Policy", 1998, Yale University, US. Connecticut. New Haven.
- "Intellectual Property Rights", 2005, University of Cambridge, UK. England. Cambridge.

"Intellectual Property Rights", 2005, Imperial College London, UK. London.

* Where technology transfer office is involved in exploitation
** Where technology transfer office is not involved in exploitation

7.5 Presenting Benchmarked Pattern of Commercial Knowledge Transfer

Based on the results obtained from the comparison methods, policies and regulations in the five universities with respect to management and commercialization of intellectual properties generated in the university, the benchmarked pattern of commercial knowledge transfer is introduced using Figure 1.

The main activity areas: The review noted that all of the five universities focus more on the Applied Sciences, such as life science, Engineering, and so on which have higher potential in commercial exploitation.

Technology transfer office: Each of the five universities has specialized units for implementing intellectual property management and commercialization affairs.

Notes about these units can be grouped as following:

Providing the structure and investment necessary to support intellectual property: these technology transfer offices provide structures for filing, protecting and licensing intellectual property and supplying expenditures for these activities.

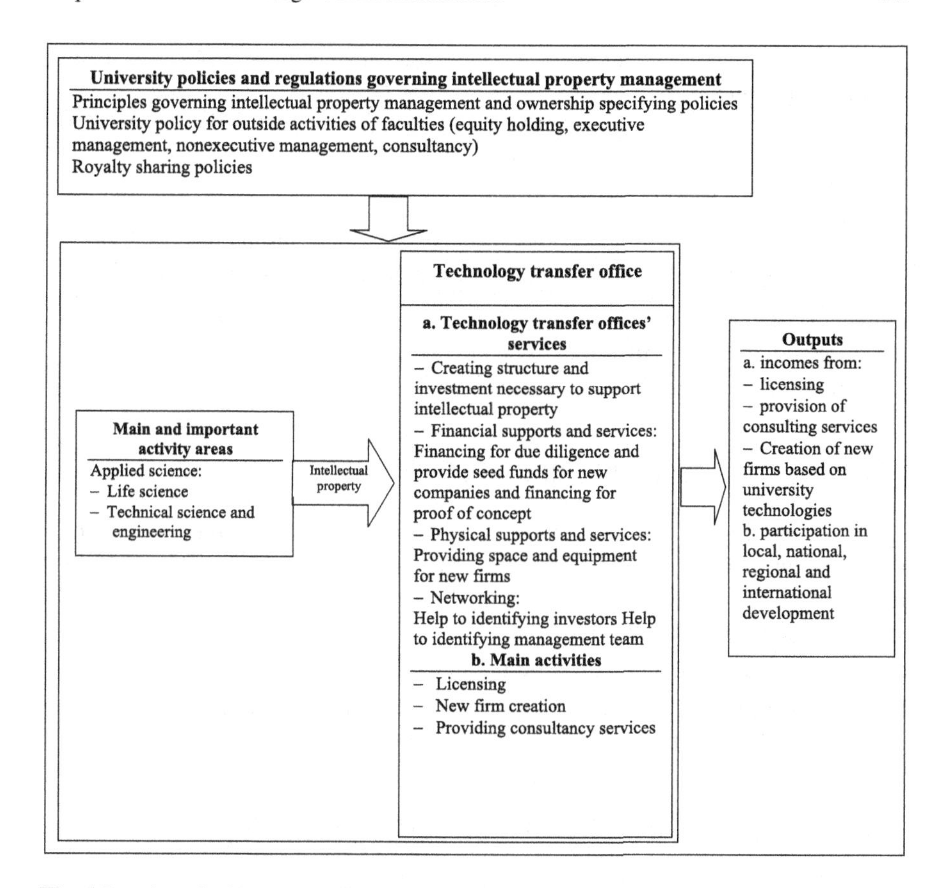

Fig. 1 Benchmarked Pattern of Commercial Knowledge Transfer

Financial supports and services: financing for a due diligence and providing seed funds for new companies as well as financing for proof of concept are among supports which are offered to researchers and inventors for knowledge commercialization.

Physical supports and services: providing space and equipment for the newly established companies are among services and supports that will facilitate and accelerate the technology commercialization.

Networking: These services usually are possible through strong and various communication and networks among potential stakeholders of knowledge commercialization.

Activities of technology transfer offices can be divided into three areas: licensing, creation of university technology-based firms, and consultancy services.

University policies and regulations governing intellectual property management: Most of the major issues considered in universities' policy related to such activities can be summarized in the following cases:

- Principles governing intellectual property management and ownership specifying policies
- University policy for outside activities of faculties (equity holding, executive management, nonexecutive management, consultancy)
- Royalty sharing policies

Outputs: finally, regarding various stakeholders, the results and outputs of knowledge commercialization can be categorized into the following categories:

a. incomes obtained from licensing, provision of consulting services and creation and management of new firms based on university technologies
b. participation in local, national, regional and international development.

7.6 Presenting the Model of Analytic Hierarchy Process

In order to review and determine the importance of factors identified in extracting pattern of knowledge commercialization, Analytic Hierarchy Process and Delphi method were used. Criteria and alternatives were identified and presented in the form of a hierarchical model:

Goal: "knowledge commercialization in universities" can be proposed as a goal.

Alternatives: most of the major requirements identified in the desired model are assumed as alternatives of the Analytic Hierarchy Process. These requirements include:

- Networking and communication between academians, industrialists, investors and entrepreneurs
- Creating the structure and investment necessary to support intellectual property
- Supplying relative freedom for faculty members to involve in business and commercial activities
- Adopting incentive policies in royalty sharing for faculties
- Providing physical supports and services (space and equipment)
- Providing financial supports and services

Criteria: The requirements of assessment criteria are categorized in hierarchical model as following:

- Effectiveness in long term
- Considering interests of all stakeholders
- Consistency with academic missions

Identifying the goal, criteria and alternatives, a hierarchical model was designed as Figure 2.

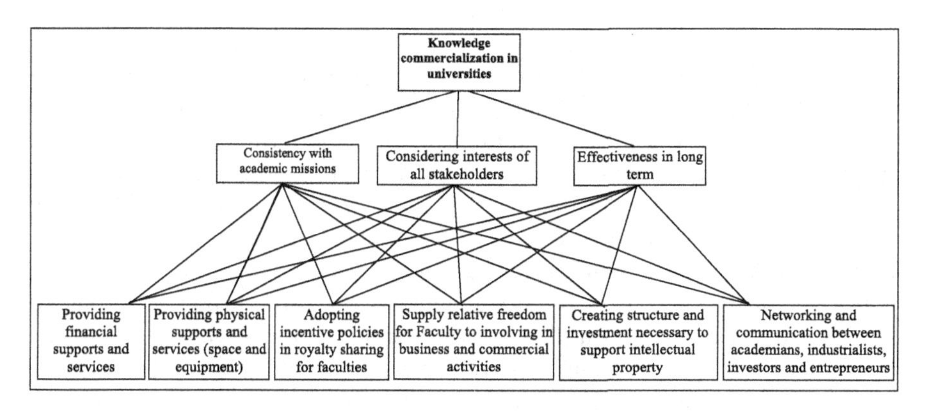

Fig. 2 Hierarchical Model of Priority Analysis

8 Results

8.1 Ranking the Knowledge Commercialization Requirements

With respect to the research method, a questionnaire was designed based on the hierarchical model for paired comparison of identified requirements and criteria (Figure 2). All matrices in a paired comparison in 24 completed questionnaires were acceptable and less than 0.1. Synthesizing the judgments of respondents in relation to priorities and preferences of the introduced criteria and requirements based on the criteria were calculated which are mentioned in appendix A.

Figure 3 shows final priority of commercialization requirement.

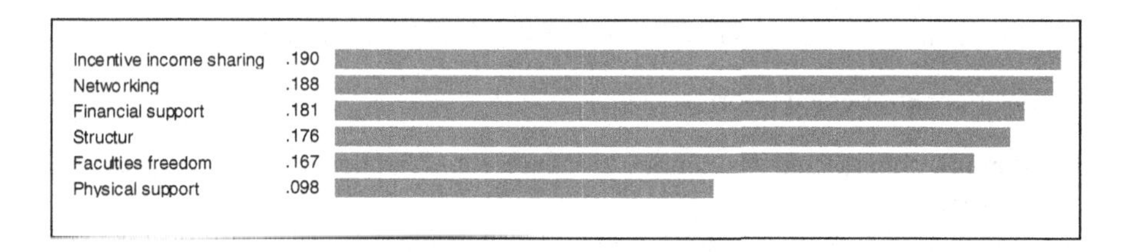

Fig. 3 Priorities with Respect to Goal: Knowledge Commercialization

9 Discussion and Conclusion

According to the findings of this study what is higher important from the perspective of academic entrepreneurship and knowledge commercialization experts is "Adopting incentive policies in royalty sharing for faculties", and "Networking", "financial support", "creating the necessary structures" and

"Faculty freedom" are placed in the next priorities, respectively. It is recommended that extensive networks for communication between the stakeholders participating in commercialization process be developed; adequate financial support and services be provided in order to facilitate the creation of firms and knowledge commercialization, and appropriate structures be created for the management and exploitation of universities' new technologies (technology transfer offices); moreover, necessary freedom should be awarded to academic researchers and faculty members to put knowledge into practice. It should be noted that in the ranking of these requirements according to the viewpoint of experts, the distance between them is not high, indicating the importance of all these requirements and necessity of paying attention to all of them.

On the one hand, the government can also provide necessary stimulus and motivation to enter into the business activities and financial independence through the reduction of financial support and allocation of budget to universities, and on other hand, it can provide suitable context for development of such activities in the university through passing facilitating and accelerating laws and policies.

In this regard, most important reforms toward development of the academic entrepreneurship and knowledge commercialization can be summarized in the two following categories:

a. University-oriented reforms:

1. Reduction of bureaucracy and creating flexibility in commercialization process.
2. Creating forums, networks and bureaus among researchers, industrialists and investors.
3. Conferences and joint meetings with industry.
4. Emphasis on the entrepreneurial activities when redefining the missions of university.
5. Implementing entrepreneurial activities for evaluation of faculties.

b. Government-oriented reforms:

1. Adopting policies and regulations to enhance communication between industry and universities.
2. Strengthening laws related to protection of intellectual property.
3. Gradually reduction of the budget allocated to universities and strengthening sensitivity and tendency toward the financial independence and leading them toward adopting measures for developing their activities in business and entrepreneurship area.

Finally, it is emphasized that the relationship of three major sectors, i.e. universities, government, and industry is defined in society area and has reflective complexity. So practical studies should be continued for better and clearer representation of this relationship in the future.

Appendix A

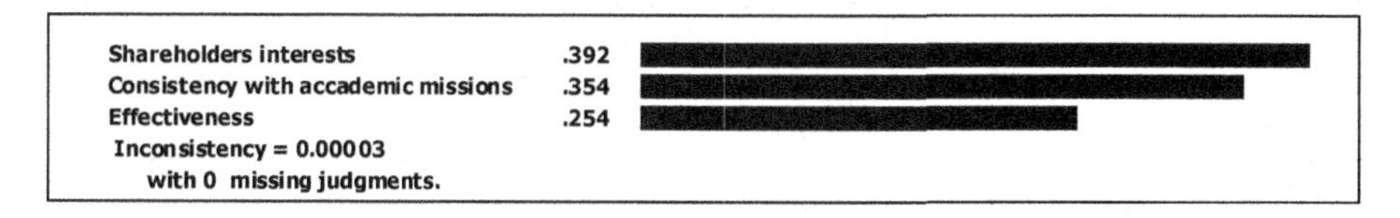

Fig. A.1 Criteria Priorities with Respect to Goal: Knowledge Commercialization

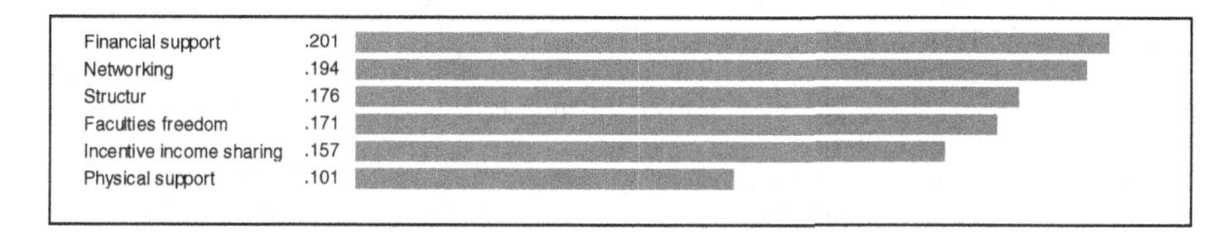

Fig. A.2 Priorities with respect to Effectiveness

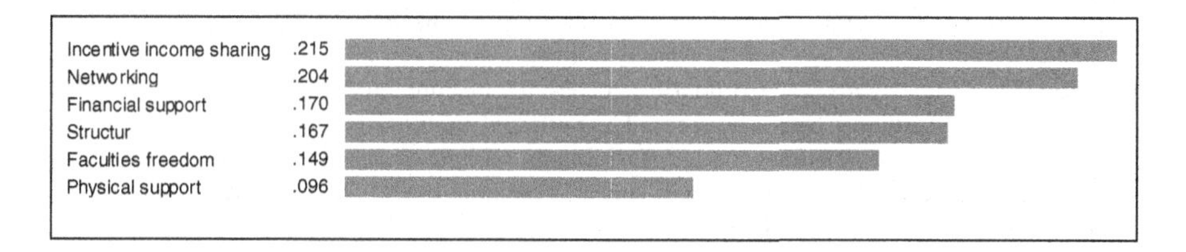

Fig. A.3 Priorities with respect to Stakeholders interests

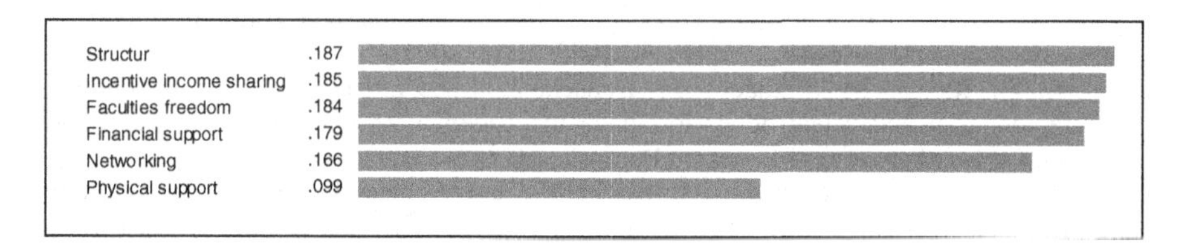

Fig. A.4 Priorities with respect to Consistency with academic missions

References

Audretsch, D.B., Bönte, W., Keilbach, M.: Entrepreneurship capital and its impact on knowledge diffusion and economic performance. Journal of Business Venturing (2008) (in press)

Baldini, N., Grimaldi, R., Sobrero, M.: Institutional changes and the commercializationof academic knowledge: A study of Italian universities' patenting activities between 1965 and 2002. Research Policy 35, 518–532 (2006)

Bozeman, B.: Technology transfer and public policy: a review of research and theory. Research Policy 29(4-5), 627–655 (2000)

Debackere, K., Veugelers, R.: The role of academic technology transfer organizations in improving industry science links. Research Policy 34, 321–342 (2005)

Decter, M., Bennett, D., Leseure, M.: University to business technology transfer-UK and USA comparisons. Technovation 27, 145–155 (2007)

Di Gregorio, D., Shane, S.: Why do some universities generate more start-ups than others? Research Policy 32, 209–227 (2003)

Etzkowitz, H.: The norms of entrepreneurial science: cognitive effects of the new university–industry linkages. Research Policy 27, 823–833 (1998)

Etzkowitz, H.: Research groups as "quasi-firms': the invention of the entrepreneurial university. Research Policy 32, 109–121 (2003)

Etzkowitz, H., Carvalho de Mello, J.M., Almeida, M.: Towards "meta-innovation" in Brazil: The evolution of the incubator and the emergence of a triple helix. Research Policy 34, 411–424 (2005)

Etzkowitz, H., Leydesdorff, L.: The dynamics of innovation: from National Systems and "Mode 2" to a Triple Helix of university–industry–government relations. Research Policy 29, 109–123 (2000)

Goldfarb, B., Henrekson, M.: Bottom-up versus top-down policies towards the commercialization of university intellectual property. Research Policy 32, 639–658 (2003)

Grimaldi, R., Grandi, A.: Business incubators and new venture creation: an assessment of incubating models. Technovation 25, 111–121 (2005)

Henrekson, M., Rosenberg, N.: Designing Efficient Institutions for Science-Based Entrepreneurship: Lesson from the US and Sweden. Journal of Technology Transfer 26, 207–231 (2001)

Klofsten, M., Jones-Evans, D.: Comparing academic entrepreneurship in Europe - The case of Sweden and Ireland. Small Business Economics 14, 299–309 (2000)

Landry, R., Amara, N., Ouimet, M.: Determinants of knowledge transfer: evidence from Canadian university researchers in natural sciences and engineering. Journal of Technology Transfer 32, 561–592 (2007)

Llano, J.A.: The university environment and academic entrepreneurship: a behavioral model for measuring environment success. Howe School of Technology Management (2006)

Lockett, A., Wright, M.: Resources, capabilities, risk capital and the creation of university spin-out companies. Research Policy 34, 1043–1057 (2005)

Looy, B.V., Ranga, M., Callaert, J., Debackere, K., Zimmermann, E.: Combining entrepreneurial and scientific performance in academia: towards a compounded and reciprocal Matthew-effect? Research Policy 33, 425–441 (2004)

O'Shea, R.P., Allen, T.J., Chevalier, A., Roche, F.: Entrepreneurial orientation, technology transfer and spinoff performance of U.S. universities. Research Policy 34, 994–1009 (2005)

OECD, Knowledge Management in the Learning Society, Paris (2000)

Rasmussen, E., Moen, Ø., Gulbrandsen, M.: Initiatives to promote commercializationof university knowledge. Technovation 26, 518–533 (2006)

Siegel, D.S., Waldman, D.A., Atwater, L.E., Link, A.N.: Commercial knowledge transfers from universities to firms: improving the effectiveness of university–industry collaboration. Journal of High Technology Management Research 14, 111–133 (2003)

The Times Higher Education, World University Rankings, U K. London (2007)

Tijssen, R.J.W.: Universities and industrially relevant science: Towards measurement models and indicators of entrepreneurial orientation. Research Policy 35, 1569–1585 (2006)

Wright, M., Clarysse, B., Mustar, P., Lockett, A.: Academic Entrepreneurship in Europe. Edward Elgar, Cheltenham and Northampton (2007)

Case Study: Innovative Methodologies for Measuring and Mitigating Network Risk in the Electricity Distribution Industry

Simon Blake[1], Philip Taylor[1], and David Miller[2]

[1] School of Engineering and Computing Sciences, Durham University, Durham DH1 3LE, United Kingdom
S.R.Blake@durham.ac.uk
[2] Northern Powergrid, Manor House, Station Road, New Penshaw, County Durham DH4 7LA, United Kingdom

1 Introduction

Northern Powergrid are responsible for the distribution of electrical energy to 3.8 million customers in the North East of England, Yorkshire and North Lincolnshire. They are accountable to the industry regulator OFGEM, in particular for the level of network risk, as measured by the frequency and duration of interruptions of supply to customers. Durham University has extensive experience of research into Power Systems, and of working with the electricity distribution industry both in the UK and overseas.

In 2006, Northern Powergrid (then called CE Electric) approached Durham University regarding collaboration on a programme of measuring and then seeking to mitigate the level of network risk. There was a concern that this level might worsen in the foreseeable future, as a consequence of factors including more sophisticated customer requirements, extreme weather events, increasing vandalism, smarter grids as part of moving to a low carbon economy, and an ageing asset base. It was decided to approach this challenge as part of a PhD research project, which was carried out by the present lead author over a 3 year period up to September 2010. This PhD succeeded in developing methodologies to measure network risk, and applying them both to straightforward and to complex case studies on the distribution network including the replacement of ageing assets, the implementation of automation to improve efficiency, larger power loads as a consequence of increasing take up of heat pumps and electric vehicles, and the challenges posed by increasing levels of distributed generation including renewables. The results of this research were published as an invited book chapter and as research papers presented at 5 different international conferences during the course of the PhD.

As the PhD neared completion, there was concern that the understanding already gained should not be lost, but should if possible be transferred to and implemented within Northern Powergrid. Various options were explored, and the most potentially effective appeared to be a Knowledge Transfer Partnership (KTP)

R.J. Howlett et al. (Eds.): *Innovation through Knowledge Transfer 2012*, SIST 18, pp. 195–198.
DOI: 10.1007/978-3-642-34219-6_21

between Durham University and Northern Powergrid, with additional funding from the Technology Strategy Board. The reasons for choosing this route included the successful experience that Durham University already possessed in setting up and managing KTP projects, including at a post-doctoral level, the potential support of the wider KTP network, and the desire to retain full involvement from the university as knowledge base, and from Northern Powergrid as the company partner. The grant application and proposal process for a 3 year KTP project took around 12 months to complete [1], and the present lead author was employed as KTP Associate from December 2010.

2 Knowledge Transfer to Date

The Project Plan has been revised once during the KTP, and now includes 7 strands running in parallel, some of which are expected to result in the development of software packages for use in Northern Powergrid both during and after completion of the 3 year KTP project. The first strand concerns the prioritisation of asset replacement. Existing condition based risk management databases contain regularly updated information on the condition of major overhead power lines, underground cables, transformers and switchgear on the network. A proportion of these needs to be refurbished or replaced each year in order to maintain a serviceable network. The cost of this activity is in excess of £100 million per year. Targeting the replacement programme where it will be most effective in reducing the overall level of network risk is vital. Algorithms have been developed to support these decisions, and they are in the process of implementation within the company. Knowledge transfer within Northern Powergrid has been achieved by frequent meetings with relevant engineers, and by the publication of several summative internal reports. A paper on the research leading to these algorithms was presented at the IET international conference on Reliability of Transmission and Distribution Networks in November 2011 [2], and it is anticipated that a further paper will be published and presented in 2012.

The second strand of the KTP concerns the prioritisation of load-based reinforcement. It is expected that the developing low carbon economy will lead to increased electrical energy demand as petrol vehicles and gas domestic heating are replaced by electric vehicles and heat pumps. This increasing demand will progressively require the uprating of circuits which could otherwise become overloaded. Identifying these circuits, and finding the most cost-effective ways of uprating them, is essential. Again, methodologies and algorithms have been developed to support these decisions, and meeting and reports have led to knowledge transfer within the company. Some of the early work in this area has already been presented in research papers at the industry-leading CIRED international conference in June 2011 [3, 4], and it is anticipated that further papers will be published and presented in late 2012 or early 2013.

The first two strands are applied to the whole distribution network. The third strand concerns the development of a project investment appraisal tool, for evaluating options for network development at individual locations. A number of case studies have been undertaken, working closely with network design engineers in each case, and the underlying methodology has been continuously developed, first as a manual algorithm, and more recently as a software package. This strand has seen growing acceptance and involvement within the company, as knowledge is shared often on a one-to-one basis. It well illustrates the progressive nature of the knowledge transfer process, with a steady if sometimes slow growth of the acceptance and use of what is at first an unfamiliar and occasionally radical approach.

The remaining strands are less developed at the time of writing. These strands are: a policy development toolkit, refining output measures, ensuring data integrity, and contributing to the low carbon networks project. The last of these is already providing useful synergies, as this £60 million project involves teams from both Durham University and Northern Powergrid. KTP experience working between these two institutions facilitates useful knowledge transfer in both directions.

3 Conclusions and Outlook

One third of the way through the 3 year KTP, it has already delivered tangible and quantifiable benefits to the company, as well as to the university and to the international academic and industrial community. Overall control of the project is exercised by the local management committee (LMC) which has already met on three occasions. At a recent review, the Company Supervisor concluded that the project was pleasingly progressing to plan. The KTP Associate had refined the network risk philosophy and agreed it with thought leaders within Northern Powergrid. From there, he has provided informative case studies and a critical review of the company's investment programme that will inform future plans. The Knowledge-Base Supervisor considered that the case studies and reports already produced were proving useful both for Northern Powergrid and for Durham University. Flexibility was required and had been shown as regards the variety of workload and the deadlines to be met. However, patience was required as regards taking ideas and methodologies and embedding them within Northern Powergrid working practices.

Although most KTPs are set up with smaller companies, the present KTP project demonstrates the very real benefits that can accrue from a KTP based at a large company which is already an established industry leader. The unique location and experience of the KTP Associate, and the full and detailed support provided from the university, the company, and the wider KTP community, have enabled ideas to be freely developed outside company structures, and yet also to be implemented effectively within these structures. It is expected that further penetration into the company, and implementation of algorithms, methodologies and associated software, both those already developed and those still being developed, will continue throughout the remaining two years of the KTP. It is further expected that both Northern Powergrid and Durham University will continue to benefit from this association.

References

1. Knowledge Transfer Partnerships, Grant Application and Proposal Form, Network Risk Project (approved August 2010)
2. Blake, S.R., Taylor, P.C., Miller, D.C., Black, M.: Incorporating Health Indices into a Composite Distribution Network Risk Model to Evaluate Asset Replacement Major Projects. In: IET Reliability of Transmission and Distribution Networks Conference, London (November 2011)
3. Blake, S.R., Taylor, P.C., Miller, D.C.: A Composite Methodology for Evaluating Network Risk. In: CIRED 21st International Conference on Electricity Distribution, Frankfurt, Germany (June 2011)
4. Blake, S.R., Taylor, P.C., Creighton, A.M.: The Value of Distributed Generation for Mitigating Network Risk. In: CIRED 21st International Conference on Electricity Distribution, Frankfurt, Germany (June 2011)

KTP and M-Commerce: Innovation in the Building Industry

Paul Crowther

Faculty of Arts, Computing, Engineering and Sciences
Sheffield Hallam University,
Howard St, Sheffield S1 1WB, UK
P.Crowther@shu.ac.uk

Abstract. In spite of practical e-commerce and m-commerce issues being regarded as solved problems by the academic community, small to medium sized enterprises are still having implementation problems. Likewise, academics are not always fully aware of the issues facing industry in terms of adoption of new (and not so new) technologies and adaptation of processes.

This paper will examine the nature of KTP's (knowledge transfer partnerships) and a case study involving an m-commerce project in the building industry. This is innovative for this sector giving the company which is the subject of the case study a potential first mover advantage. Factors to be considered when developing contemporary e-commerce projects and the challenges of m-commerce will be discussed.

1 Introduction

Small to medium sized enterprises (SME's) are defined by the European Commission as enterprises employing between 50 and 250 employees. (European Commission: Enterprise and Industry, 2011). It is not uncommon for SME's to have a web presence of some type. This can range from a simple web page giving contact details through to a fully functional site which allows interactions such as allowing customers to purchase products, purchase after sales service, receive news and contribute to discussions.

Most SME's generally find it difficult to develop an e-commerce system in house because few employ a dedicated information technology support officer. In the few cases where they do, that person is often fully occupied maintaining and supporting the companies core processes. Because of this lack of expertise or time they normally take one of two options:

- Outsource development
- A gifted amateur in the company develops a web presence

Typical problems that arise in both cases include

- poor documentation
- a static website which is difficult to update

R.J. Howlett et al. (Eds.): *Innovation through Knowledge Transfer 2012*, SIST 18, pp. 199–205.
DOI: 10.1007/978-3-642-34219-6_22

- no backup policy
- data inconsistency and redundancy

If e-commerce is regarded as a problem by SME's, then m-commerce is often considered too hard and costly, if indeed, it is considered at all.

This corresponds with the inhibitors identified in ATG's (2010) white paper: costs, lack of IT capacity and lack of skills and knowledge. These kinds of problems have led to a certain disillusionment with e-commerce within small to medium sized enterprises whose web presence is often little more than basic information and a contact details.

The case study presented here is based on the experiences of a knowledge transfer partnership between Sheffield Hallam University and Grayson Fittings and Fixtures. Grayson is part of a larger group of related companies in the construction industry. It supplies materials to other the members of the group and also to other unrelated companies. It is primarily interested in business to business (B2B) transactions.

Before the start of the project, Grayson faced the following problems:

- A web site with no online catalogue and no sales capability
- No integrated update policy
- Problems with an IT software supplier

2 Knowledge Transfer in Knowledge Transfer Partnerships

Knowledge Transfer Partnerships (KTP's) are partnerships funded by the government to facilitate the transfer of knowledge between academia and business. This is by no means a one way knowledge transfer process.

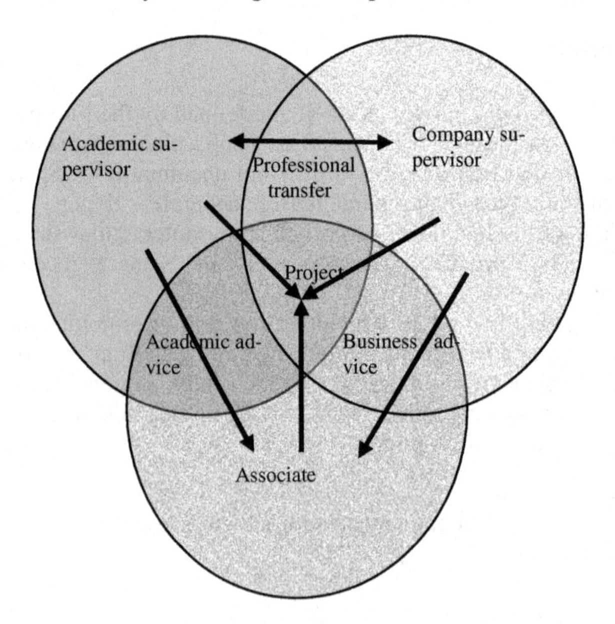

Fig. 1 Knowledge transfer between parties in a KTP (Crowther 2011)

The knowledge transfer partnership scheme involves at least three individuals as shown in figure 1. The academic and company supervisor guide the project with the company supervisor providing specific company knowledge. The associate is employed jointly by the university and the company and is usually a graduate with skills in the subject domain.

The academic supervisor could be regarded as a consultant. Where there is an external consultant working with an organisation, Wang et al (2007) suggested knowledge transfer was a function of the competence of the knowledge holder and the absorptive capacity of the organisation. The Wang et al's model was not a good fit because of the nature of the KTP.

The model described in figure 1 better describes the situation in this case study. The difference may be explained by the nature of the consultation and the case study being based in an SME whereas Wang's study concentrated on large organisations and a different form of consultation. The knowledge transfer was more in line with Cook and Brown's (1999) 'generative dance between knowledge and knowing'. Figure 1 illustrates this where there is professional transfer shown between the academic and the company supervisor. Here "... the source of new knowledge and knowing lies in the use of knowledge as a tool of knowing within situated interaction with the social and physical world." (Cook and Brown, 1999, p. 383).

Interaction takes place rather than knowledge transfer, so rather than the distinction between tacit and explicit knowledge, there is the concept of knowing. This knowing is the result of interaction between the team members. In the case study, all members of the KTP team contributed to the group knowledge.

The interaction also provides a trigger for innovation. 'Innovation involves the conversion of new knowledge into a new product, process or service and the putting of this new product, process or service into use' (Johnson et al (2011), pg. 269). In this case study the development of a mobile commerce presence in the building industry is an innovative development directly facilitated by knowledge transfer.

3 Case Study

The project described in this case study was aimed at the business to business (B2B) market, however there was nothing to stop individuals purchasing items (B2C or business to consumer). In terms of B2B, Grayson was in the one-to-many private e-market and transactions were in the form of direct sales (Turban et al, 2006). Customer relations are linked to sales as the building supply industry is highly competitive. Generating repeat business and repeat visits to the company web site was important.

The project was the third attempt by Grayson to develop an e-commerce system. The earlier two attempts involved a contractor and a 'gifted amateur' and neither were successful. As a result there was a web presence, but little else. Sales were predominantly made over the telephone, although this was a little easier than in some other sectors because of the standard terminology used in the building industry.

In this case the development of the system required a major analysis of the data stored and the data structures involved in the existing inventory system to create a hierarchical on-line catalogue. A second objective was to make the site usable in a mobile environment. An analysis of traffic at another company in the group which has a functional e-commerce system revealed only a small percentage of site accesses came from mobile devices (2.2%) although the site was not configured for mobile devices. By far the biggest platform was Microsoft Windows (88.8%), followed by Macintosh (6.5%), then iPhone (1.1%). Although the m-commerce component of activity is currently low, it was considered important to develop it because of the increasing prevalence of smart phones and the advantages of having multiple channels or contact points for sales (Baird and Rowen, 2011).

In developing the system, the first task was to re-engineer the indexing system of the interface with the inventory control system. This was needed to optimise searches by users of the on-line system. Products were categorised for easy visual searching (figure 2).

Consideration was given to hosting the searchable catalogue using cloud technology. Testing revealed issues with update and retrieval time leading to time out issues. Although it was believed these could be solved, it was decided to continue with conventional hosting for the deployment.

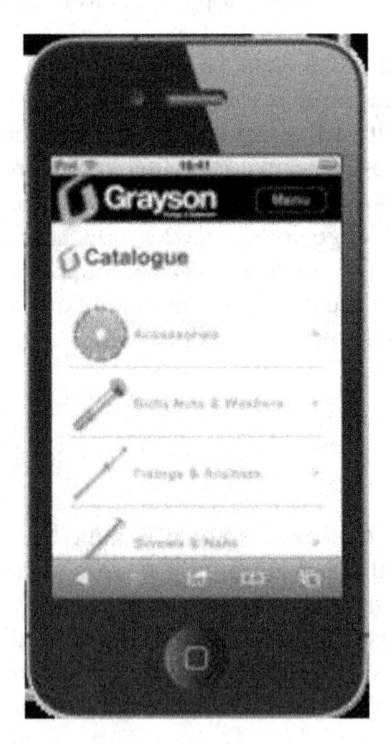

Fig. 2 Simulation of the m-commerce site

Development of the mobile version was concentrated on Apples i-phone. Initial development work was completed using a simulator. This was then tested on an ipod touch bought specifically for the project. During alpha testing personal i-phones were used.

The project was completed to the satisfaction of the stakeholders at Grayson. The on-line catalogue was working (http://www.grayson-gb.com/Products) and had gone live after extensive alpha testing, also known as acceptance testing phase (Sommerville, 2001). In the first week of operation form 8th August 2011 there was a 77% surge in traffic to the site as measured by Google analytics. Prototypes of the mobile version have been developed and are in the beta-test phase with trusted clients. Different methods of doing the sales on-line and different hosting options are being evaluated.

Development of the e and m-commerce sites was a primary objective of the project, but an equally important facet was the ease of maintenance. Maintenance issues were not rated highly in Addison's (2003) survey where it was 19th out of 28 ranked risk factors (although academics ranked it 12th). Clients/Users and Developers saw it as a much lower risk, but this was a relatively early study. Grayson is continually evolving its product lines, therefore the marketing section needed to be involved in the sites update. This was specifically included in the design brief because of earlier experiences and continually stressed by the academic partner.

4 Discussion

Chaffey (2007) summarised the attitudes to business benefits of online technologies. The case study conformed to 5 out of the top 10 factors, namely:

- reduce costs/increase efficiency/profit
- Improve communication and relationship with customers
- Keep up with competitors
- Increase speed of access to information
- Increase customer base

Chaffey also detailed tangible and intangible benefits from e-commerce and e-business and these also mostly conformed to what was observed in the case study. Among the intangible benefits were:

- Corporate Image Communication
- Enhancement of Brand
- More rapid, responsive marketing
- Improved customer service
- Feedback from customers on products

In the building industry, materials are often ordered from the building site and rapid, responsive marketing is essential to maintain a competitive edge. The m-commerce implementation was designed to address this issue. Research of Grayson's competitors websites suggested this was an innovate approach potentially giving them a first mover advantage. It is too early to evaluate if these benefits have been realised. However they are among the anticipated benefits management are expecting.

An important factor in the development of the e-commerce project was that of website quality, specifically navigability and security. Lee and Kozar (2006) found this to be the most important consideration of on-line customers. Surprisingly, their study found service quality was relatively unimportant (but they noted this required more investigation). Navigability was a concern for the m-commerce version of Greyson's site to make sure customers could easily find the product they were searching for on a small screen format.

Sung (2006) discusses critical success factors (CSFs) for e-commerce. This was a comparison between Eastern and Western countries (with some limitations), but 4 CSFs were in common:

- customer orientation,
- ease of use,
- delivery of goods and services
- variety of goods and services

In terms of the case study, all but the last of these points were addressed. The last point was problematic because of the specialist markets the company operates in.

5 Conclusion

Technologies which are considered mainstream by academics are often innovative when applied in a sector for the first time. This was the case with this case study where the organisation gained first mover advantage in the building supply sector as a direct result of knowledge transfer. The role of the knowledge transfer partnership in this case was for the academic partner to bring the necessary technical and development skills and apply them in the context of the businesses requirements. As a result of this the partnership resulted in an innovative solution, although the component parts were not individually innovative.

The success of the project was the result of adhering to the following methodology. When considering developing a e-commerce or e-business presence (or indeed m-commerce), a number of issues should be considered along with the normal cost benefit analysis which should be undertaken before any systems development:

- what is the primary aim
- who are the main users
- how will the site be accessed
- how will the site be maintained

The primary aim of the move into e-commerce was to increase volume and efficiency of sales, however, an important second objective was to improve customer relationship management. Developing an easier way for customers on a building site to order materials was part of this. A quick survey revealed no competitors had made the move to m-commerce despite the fact most customers had access to some sort of smart phone. The partnership considered that mobile access was going to be important and adaptive interfaces would be essential to minimise maintenance and deploy on as many types of interface as possible.

Maintenance of the site was paramount and that maintenance would be in house. Meeting this requirement was essential because products, prices, information and specials are all part of today's commercial world and all change. The company could not justify having an IT specialist and wanted to keep outsourcing costs to a minimum. Training of staff was therefore essential along with periodic reviews to monitor navigability and security (to maintain website quality).

The case study needs to be revisited in the future to evaluate the success of the e-commerce systems in terms of contribution to company revenue, user acceptance and maintenance record.

References

Addison, T.: E-commerce project development risks: evidence from a Delphi survey. International Journal of Information Management 23, 25–40 (2003)

ATG (Art Technology Group), e-commerce: A guide for small and Medium Enterprises (2010), white paper available at `http://www.adeogroup.co.uk/blog/wp-content/uploads/2010/11/SME-Guide-to-ecommerce.pdf` (last accessed June 30, 2011)

Baird, N., Rowen, S.: After the Storm: Connecting with the New Online Consumer (2011), benchmark report available at `http://www.atg.com/resource.../AR-After-the-Storm-RSR-Benchmark-Report.pdf` (last accessed June 30, 2011)

Chaffey, D.: E-Business and E-Commerce Management, 3rd edn. Prentice Hall, Harlow (2007)

Cook, S.D.N., Brown, J.S.: Bridging Epistemologies: The Generative Dance Between Organizational Knowledge and Organizational Knowing. Organization Science 10(4), 381–400 (1999)

European Commission: Enterprise and Industry, Small and medium-sized enterprises (SMEs) SME Definition (2011), `http://ec.europa.eu/enterprise/policies/sme/facts-figures-analysis/sme-definition/index_en.html` (last accessed November 22, 2011)

Crowther, P.: Knowledge Transfer Partnership - A Case Study from the Building Industry. In: Proceedings of 11th International Conference on Knowledge, Culture and Change in Organisations, Madrid (June 2011)

Johnson, G., Whittington, R., Scholes, K.: Exploring Strategy, 9th edn. Prentice Hall, Harlow (2011)

Lee, Y., Kozar, K.A.: Investigating the effect of website quality on e-business success: An analytic hierarchy process (AHP) approach. Decision Support Systems 42, 1381–1401 (2006)

Sommerville, I.: Software Engineering, 6th edn. Addison-Wesley, Harlow (2001)

Sung, T.K.: E-commerce critical success factors: East vs. West. Technological Forecasting and Social Change 73, 1161–1177 (2004)

Turban, E., King, D., Viehland, D., Lee, J.: Electronic Commerce 2006, A Managerial Perspective. Pearson Prentice Hall, New Jersey (2006)

Wang, E.T.G., Lin, C.C., Jiang, J.J., Klein, G.: Improving enterprise resource planning (ERP) fit to organizational process through knowledge transfer. International Journal of Information Management 27, 200–212 (2007)

Knowledge for Business (K4B): A University – Business Knowledge Transfer Collaboration Framework

Phil Fiddaman, Robert Howie, Dominic Bellamy, and Claire Higgitt

The Knowledge Transfer Team, University of Hertfordshire, Hatfield, Hertfordshire AL10 9AB
ktp@herts.ac.uk

Abstract. Knowledge for Business (K4B) is a new knowledge transfer delivery framework developed within the University of Hertfordshire to supply a KT solution where traditional grant funded systems may not apply or prove too inflexible. It is designed to address two key drivers; a) for the University to demonstrate the impact of its academic activities and b) address a business demand to innovate by acquiring new capabilities in a highly flexible and structured way.

1 Introduction

The framework model represented in K4B, whilst new as a package to potential clients, is born out of the collective field experience of the KT Team over several years. The timing reflects the needs of all parties involved in the process given acute focus by the economic challenges we currently face.

The primary aims are:

- Introduce new commercial income from companies not qualifying for grant aided support.
- Build and nurture business relationships through K4B for longer term collaborations e.g. KTP/sKTP
- Identify the impact of academic research in the economy for the Research Excellence Framework (REF).

2 Background

The KT Team recognised a clear gap between genuine business needs for KT solutions across all types of organisation, especially SME, and grant funded packages, in particular KTP. The 2010 Government spending review exaggerated this gap with a 50% funding cut and tightening of the qualification criteria for KTP; strongly emphasising innovation, impact and challenge. Not new, but remaining largely unaddressed, was the on-going demand from business for a

R.J. Howlett et al. (Eds.): *Innovation through Knowledge Transfer 2012*, SIST 18, pp. 207–210.
DOI: 10.1007/978-3-642-34219-6_23

flexible and structured mechanism for KT solutions delivered in commercial timescales and which meet the business need. Parallel to this was an on-going need from academics for a framework in which projects could be delivered compatible with their departmental processes and need for impact recognition.

3 Research and Rationale

The KT team conducted comprehensive research into the strengths and opportunities for Knowledge Transfer projects across 6 selected Schools of study. This involved Heads of School and their Business Development Teams.

Reasons from clients for not engaging in KTP/sKTP were recorded in terms of funders' eligibility criteria, project profiles and protracted start up times for urgent and tactical projects. Consultancy projects were identified as problematic in terms of time allocation, business timing expectations and relatively high costs. Research into the SME sector identified many projects important for them but which were not eligible for grant funded schemes and hence not benefiting from University engagement.

Findings concluded the most suitable form of non-grant funded business engagement with the University was through a mentoring/teaching format recognising academic availability with project spreads of 6 to 9 months. This was re-enforced at The Enterprising Academics Seminar Conference in 2011 where ex-Dragon's Den entrepreneur Doug Richards advocated a mantra for universities: "Teach don't Consult"

4 Results

The K4B model emerged. Academic staff are timetabled efficiently for a few hours per week to mentor existing company staff with a pre-defined business project to disseminate new knowledge into an organisation. The plan has clear tasks and deliverables with given time scales. Project delivery is monitored and mutually signed off monthly to maintain planned progress. Flexibility allows for project change or adaptation including additional academic disciplines where required. The K4B project is underwritten by a collaboration agreement and fully funded by the client.

Six projects were successfully launched in a 2011 pilot campaign valued at £77K. Projects include – On-line business development, digital printing, materials testing and strategic marketing.

From experience to date, K4B has evolved to include the appointment of a University graduate intern – K4Bi - where the client does not have appropriate existing skilled staff enabling the delivery of more sophisticated projects and potential employment opportunities. In addition, through support funding from the University's HEIF budget we are offering the scheme at subsidised rates to the third sector, under the brand Knowledge for Society (K4S).

5 Discussion

K4B is proving to be an enabling framework for project work originating from and compatible with broader objectives and plans. Examples:

- A bridging project between a sKTP and start of a KTP where the company is keen to maintain momentum.
- Preparing and scoping for larger project applications e.g. KTP
- Product development project alongside a live iCase research output.
- Feeding into current PhD projects
- Creating graduate employment opportunities
- Feedback into teaching.

6 Conclusions

K4B is proving to be a viable commercial income generator readily embraced by business and academics. The fusion of an easily understood structural framework and the ability to work flexibly with existing institutional processes is set to become a popular brand for the University and the local business community.

References

Exchange Communications Ltd, St Albans – "We were particularly attracted to the knowledge exchange product K4B; it struck us as tailor made for our business requirements. The scheme easily coped with any changes required. We can see some real benefits coming through our new online sales lead generation married to the new social media strategy implemented." ***Dave McKinnon, Managing Director***

Alchemy Metals, Stevenage – "We are using the K4B project to help identify new European markets, markets that will allow us to utilize the full capacity of our new site. Working with our academic partners has been rewarding and has changed our opinions of academics; we now know they don't all live in 'ivory towers'. We are all very excited about the project, and already have some early wins from the resulting data." ***Karen Greasby, Director***

Reflections on being a Knowledge Leader/Mentor - "I jumped at the chance to take on the role as business mentor/trainer on K4B projects to work with talented business owner-managers, on projects with medium/ long-term outcomes plus socio-economic impact on their business model . I am no different to many academics in that I look for effective ways to use my existing knowledge and expertise to facilitate change, being able to apply it in businesses, as well as in the classroom is a win-win situation. I am a learner, someone wedded to the idea of being a life-long learner - relishing the opportunity to learn from the business owner-manager too. This knowledge exchange process is key to the future success of universities co-creating value in their local communities. I worked on four K4B's last year, and expect to work on six more this year; do they pan out the way they were originally planned, NO. Yet, hand on heart; I can say that the outcomes and impact of these has been significant.

Business-university collaboration needs to move away from the premise that it is up to the businesses to come to universities and clearly state their problems; most businesses are unaware of their problems - they know the challenges they have! It's up to universities to go into the local communities and start the discussion and engagement, helping business owner-managers identify the problem, and then leading them towards a sustainable solution. That's a challenge for both businesses and universities, because we have not operated that way before. The K4B scheme is perfect for this, and we have a real opportunity to build on the successes and to extend it further." ***Dr Christopher J Brown, Business School, Knowledge Transfer Leader, University of Hertfordshire***

Industrial Funding Path Analysis in the Italian University System

Mariafebronia Sciacca

University of Turin, C.so Unione Sovietica 218/bis, 10134 Turin, Italy
Sciacca@econ.unito.it

Abstract. This article attempts to describe recent paths in University-Industry linkages in the Italian University system. In the last decades University vision has fundamentally changed together with its structure and goals. A large debate has raised towards and main findings have been shared around several aspects that are impossible to be adequately covered, nevertheless we present an outlook of the general trends and literature. One of the most challenging development of University has been the mass education expansion (Trow, 1973), and the rising number of undergraduates that have stimulated a sensitive growth in number and size of many Universities. In the international frame also institutions differentiation of the overall higher education system has been pointed out as a strategic focus on which the actors will be engaged in the future (Bonaccorsi 2007). Moreover, a large part of the literature has been dedicated to knowledge society (Etzkovitz, 2003) and service market economy, an issue that has incentivized an increasing interaction between University and the external actors, and its "openness" (Slaughter and Leslie, 1997). Beyond the first and the second mission (respectively represented by teaching and research), a third stream of activities has come up as "dissemination or outreach activities" (Gulbrandsen et al. 2007), exploring the degree of entrepreneurship in the current University system (Etzkovitz, 2000). University-industry knowledge transfer represents a key research subject in the economics studies, and a critical issue in science and technology policy agenda of several countries, and inevitably linked to innovation policy, whereas innovation deals not only with specific firms but with a huge field of institutions that aim to develop technological system at national level (Lundvall, 1992).

Such complex framework offers a new vision that shifts from "old University ivory tower system" towards the openness to external actors and the market has raised, even though there is no evidence that all the external interactions between university and environment are market driven (Olsen, 2007). On one hand, new needs come from the market and a growing competition for resources between universities seems to be evident, on the other hand industry is looking for knowledge-intensive fonts, a trend accompanied by a substantial reduction of internal R&D activity.

With regard to the impact of non-public funding, in the University financial data three are the main sources of funds: Students fee, fund-raising and consultancy and contracts. Even though non-public resource is not the most significant and accounts for a small part of entire University budget, nevertheless many changes have occurred in the last years, after declining of public funds.

R.J. Howlett et al. (Eds.): *Innovation through Knowledge Transfer 2012*, SIST 18, pp. 211–226.
DOI: 10.1007/978-3-642-34219-6_24

Large debate has raised questions on motivation for academics to engage with industry, on type of contracts design for industrial collaboration, on the nature of the collaborations and main barriers (Bruneel et al. 2010), on the proximity between Industry and University (Arundel and Geuna, 2004) and joint research collaboration (D'Este et. al, 2010), while few attention has been dedicated to the relation which links Industrial funding, scientific area and quality indicators in the Universities Model in the recent years, with particular reference to Italian system.

1 Changing Strategy – A General Overview

This paper investigates University and Industry recent paths and describe the relations which links Industrial funding towards performance indicators in the Universities Model in the recent years, in the Italian system, that have guided the development of new strategies linked to organization behavior, to new funding paths and to a novel governance. Our results show that a good University performance may facilitate an external collaborations, but definitively it can foster mutual benefit in University-Industry frame of relations.

The next part is organized as follows: we start with the description of the background of research activities and Industrial funds relations in European and Italian frame; we then illustrate the methodology used to collect data in an original dataset and the Research questions; we finally design a description of the main trends of the analysis and finally report the statistic results and the conclusions.

2 Background

2.1 *Focus on EU*

In most of European countries the reduction of Government funds has raised concerns about the financial stability of Universities in terms of higher education system (Herbst, 2007). Somewhere, it has caused a remarkable raising of student taxation, in some other it has been addressed a sensitive cutting policy in public funds devoted to research. So far, no economy of scale and scope has been demonstrated for Universities. Hence, the growing need of resources has incentivized Universities in capturing new resources. Furthermore, Universities have switched from a centralized model to an autonomous, self-regulated model and many studies have described the "new public management" thinking (Bleiklie, 1994) comparing the university system to a service enterprise embedded (Olsen, 2007). This process has been encouraged by the increasing autonomy, in terms of reallocation of resources and accompanied by necessity to identify an evaluation protocol of their performance. In France and Germany[1] lump sum budgeting trend has permitted a wider discretional use in the resource allocation for Universities. In parallel, the development of an evaluation system has been set up in other

[1] In the Germany case funding of Science were distributed on a competition between public and private (Drittmittel).

countries. This process is generally based on historical input/output model, as for example the UK experience of RAE and REF and other similar practices that have occurred in other countries as France Spain (PNICDT[2]) and Holland[3]. Besides the general trend of evaluation of teaching and research activities that links the funding model to other parameters, there is a third stream of changing rules in the allocation of resources driven from historical–base to performance evaluation, deeply influenced by the decline of Government funds income (Palumbo 1999, Grossman and Helpman 1991, Salmi Hautman 2006). Large part of literature has argued that evaluation and performance parameters could affect reputation and in a certain indirect way Industry funding trends.

2.2 *Italian University System*

Before going in depth into research details some premises have to be pointed out on structural conditions of Italian system, and its main characteristics, with reference to the decentralization trend and the evaluation process.

First, as happened for European countries Italian universities have registered a significant drop in government funds, estimated in a -5% in 2000-09; the age of Professors is definitely over the EU relative colleagues while the salary is under the EU average; many authors assert that transparency in the selection process is not still guaranteed in the current system, even though decision policy should have broken with traditional system.

Second, looking at the self regulation trend, this process has started in 1989 with law 168, which enhanced autonomy and self-regulation from administrative perspective; further steps towards higher flexibility of local Universities have been made and these change have generated an higher discretional power to attract external funds and have increased University governance autonomy to set rules at local level. On the other hand, although the abovementioned premises, this process has not followed a linear trend in the following years 2000-2010. In fact, further government policies have not been substantially changed the University decisional process with regards to the organizational structure and to the funding management, so that many authors believe that gradual and conservative attitude and typical negotiation style still persist. In particular, the critical leverages of selection procedures and remuneration remained under the control of the government through the Ministry of University and Education (MIUR) (Fini and Grimaldi, 2011). Nevertheless, cut of Government funds has been steady and at the same time differentiated within universities, partially driven by Universities performance (Capano, 2011). From the HEIRD point of view, Bologna process in EHEA edge[4] has introduced a series of innovative aims throughout the last decade, but it had few direct consequences in the funding model (Moscati, 2010).

---

[2] Plan nacional de investigaciòn cientifica y desarrollo tecnològico is a pluriennal plans for the definition of resources addressed to research.

[3] National Counseil of Research in Holland has rivendicated the trend of a funding distribution based on historical data (Nwo).

[4] European Higher Education Area, an initiavite launched in 2010 aims to ensure more comparable, compatible and coherent systems of higher education in Europe.

These legislative drivers together with the development of different external (changing environment) and internal (increasing number of students) have caused the revision of a University strategy and on the other side a different approach towards the attractiveness of new external sources of funding. For this reason, in our model we consider the evaluation process as one of the significant variable that link government and industrial funding.

Thus, although some attention has been devoted to private participation in the university funding issue from literature, and the main findings are that Industry funding are characterized by an high level of competition and by a continuous short-term evaluation of research outputs; Industry funding tends to be funneled towards top universities; subsidy model has changed at the European level (Geuna 2001), few attention has been dedicated to the relation which links Industrial funding, scientific area and quality indicators in the Universities Model in the recent years, with particular reference to Italian system.

The scope of this study is to relate industrial funds and users that generate and manage U-I relations, and looking at the historical development of these connections.

We associate Industry funding and performance of Universities on a panel data representing the quasi-totality of the Italian Universities. We base the analysis starting from the assumptions derived by the literature with the integration other intuitions coming from empirical analysis with some caveats.

First of all, through the definition of efficiency as the capability to make the best use of input in order to generate the best outputs, in other words making reference to the technical efficiency, which relies on how resources are employed in a non-market system as University is (Bonaccorsi et al., 2007).

Secondly, beyond the innovative approach we focus on research inputs and outputs, consequently our indexes include also physical parameters as composition of staff, presence of scholarships and PhD over graduates. Universities have a multi-inputs and multi-outputs production (Bonaccorsi, 2007), a caveat is represented by the fact that studying such a complex frame is clearly difficult but it is not the aim of this work, moreover we are interested in providing a frame on U-I relations trend. Thirdly, studying industrial funding as mentioned above is equal to analyze a small subset of the entire budget, for this reason we undertake some statistical methods to adequately represent also small numbers. Fourthly, we exploited different control variables as size, geography and type of education to assess and clarify the influence of fixed effects on the regression variables.

3 Methods

In this paragraph, we illustrate the approach and the methodology employed to define the empirical models. We primarily collected financial and qualitative data in an original dataset and gathered information through three different approaches. First we refer to MIUR database for the financial data; then we gathered data from Nuclei Miur Evaluation; thirdly we adopted PRO3 Cineca performance indexes according to the Institutional perspective and aims.

We observed variables on a panel data collected from 2006 to 2009 due to several reasons. Since all financial data have been significantly transformed from 2006 on, and few comparability and feasibility could be assigned to previous data we decided to exclude the years before. Second, since the availability of 2010 data is partial and performance evaluation is partially covered, we dropped 2010 observations. Thirdly, the aim of this research is to consider a balanced sample including the same universities along the mentioned period.

Our approach consisted initially in the collection of a sample of 75 Italian universities. We have cleared all Universities reporting an high range of missing values in the financial data, and we finally obtained a sample of 69 Universities. Two universities have been finally dropped from regression since they presented only zero value in industrial indicators, probably due to incoherent data classification. We have referred to Nuclei MIUR Cineca dataset[5] to be able to identify on one hand economic data linked to scientific research, and on the other hand to obtain the department level data, that let us distinguish scientific and non scientific area. Finally, we have merged these information with other performance indicators partly built by an institutional program for establishment of the University aims (Pro3[6]), and party inspired by the literature[7].

Then, we have investigate the distribution and the best performing Universities throughout indexes originated in order to recognize degree of autonomy and attractiveness for Universities. The analysis is consequently based on performance indicators, aiming to describe attractiveness through panel data will be shown at University level. The last part of the analysis is focused on the relations between evaluation indexes and other financial indicators. Time lag for the model estimation which takes into consideration VTR ranking scores is due to the publication of these indexes, which took place with a three years lag; thus evaluation of 2001-03, published in 2006 has been assumed as having its effect in years 2006-2007 and in the long run in 2008 (M5).

H1) Is the degree of autonomy associated to higher efficiency? Is the industrial funding linked to efficiency or international relations?

We assume degree of autonomy as the ratio of non government funds over total income in research. In this ratio are not only counted private subjects but also other public parties different from government. We also repeat the estimation considering private funds, but as dependent variable we put both performance measured as professor over total income and degree of international attractiveness (both indexes are inspired and partially changes to evaluation process driven by MIUR[8]).

---

[5] Online reference is Nuclei.miur.it/sommario, the data enclose different set of information grouped by fareas.

[6] Programmazione-triennale.cineca.it, Pro3 is a program of data collection launched by MIUR in collaboration with CINECA consortium (Art. 1-ter of Legislative Decree 31 January 2005, No. 7, into law March 31, 2005, n.43), and has oriented the definition of the pluriennal aims of Italian universities based on qualitative and quantitative data.

[7] See Bonaccorsi et al. 2007 for a more detailed description.

[8] See note 7.

H2) Is Government funding decline a determinant in to industry funding?

The rationale here is to estimate the variation of government funds respect to the variation of industrial funds. We expect a different degree in the universities, due to different extent of collaboration. In this frame the best case should react more to a decrease of public funds. Size could affect substantially all these variation . We also control the incidence of size, north and type of University.

H3) Are Research oriented Universities performing better in terms of funding attractiveness?

In the assessed model we try to input research orientation measured as the average of several inputs. This index includes Number of PhD Scholarship financed by external funds over the total Number of scholarship; Number of Scholarship per Doctoral course, Economic availability per professor. The higher the score is, the more Research oriented the university are.

H4) Is quality evaluation relevant for Industrial attractiveness?

Last research question concerns the quality evaluation index VTR[9] developed by MIUR in the 2001-2003. In our frame, it represents a control variable, able to extend all previous question research, taking into account a possible ranking of Universities. This approach presents some caveats. First of all, time lag could be seen as a limit of research, but in our assumption since this data have been published in 2006, they could determine some influences from this year on, that corresponds to our time analysis. Second, even though there is a single quality index per University, we suppose that University ranking is not volatile and its effect persist for a certain period. This is also supported by the fact of a complex evaluation as VTR has consisted in.

4 Definition of Variables

As mentioned above, we assume the degree of autonomy as a the share of non government funds over total income for research, while Industrial funds over total income from research activities is the dependent variable, and aims to measure the degree of private funds.

Several statistics concerns have raised over this subject due to multidimensionality, presence of outliers. In our study we mitigate scale-effect measuring each component on the total income deriving from research activities, and look for a correlation between the decreasing of Government funds and industrial paths; we also look at other sources of funds as International contracts, no-profit funds.

---

[9] VTR 2006 (Valutazione Triennale della Ricerca) is an evaluation process carried out and managed by the Committee for Research Evaluation (CIVR), art. 5 Legislative Decree no. 204 on June 5th, 1998 and following modifications and integrations. It is organized in areas courresponding to 14 scientific-disciplinary Area of the National University Committee (CUN). For a detailed description see: http://vtr2006.cineca.it/documenti/DM2206_EN.pdf

We expand the relation with industrial funds considering also structural and physical variables of each university (H1 e H2), to understand the determinants that foster the collaboration with reference to a greater degree of autonomy and Government decrease. In H3 we adopt variables that may represent research orientation of the university and in H4 we add performance and quality indicators to universities and in a nested model to scientific Departments only.

Then, we try to consider only scientific area to see if estimation improve. Independent variables are performance indexes built on staff engagement, research orientation and on the other side Government funds decrease. In all the model we adopt analysis of variance to test the robustness of our models. A set of control variables have been used to limit some university-fixed effects.

Size

The foundation is to control the great variability that could affect the model: size has been calculated by using the Institutional classification of MIUR, set by the Number of students (both undergraduate and graduate), resulted in four classes (micro, mini, medium, mega). Our rationale behind this hypothesis is a better performance in the center classes, since in the mega class we expect a dispersion of economies of scale and in the smallest we do not expect economies of scale.

Type of University

In the Italian system four classes have been identified: Public, Private, Advance School of Doctorate and University for Foreign studies. In our data sample we considered the total of Public Universities and excluded some of private and University of Foreign studies due to missing data. With regards to this hypothesis we aim to understand whether private and Advanced School of Doctorate could better perform in industrial relations.

Polytechnic

The polytechnic Universities have been identified as a dummy variable and we expect that it might likely to set up an higher extent of technology skills towards industry. For this reason a polytechnic is more reliant to collaborate with firms, even its typical small size.

Geography

Most of literature suggest a slight variance across regions (Baldini , Grimaldi). Even this is not the aim of this paper we also take into consideration North as a dummy variable to define the extent of its effects.

General University

Defining general university versus specialized universities means distinguish universities offering both scientific and non scientific knowledge. Consequently in research activities from those that have 1 to 5 departments focused on specific matters (both only scientific or only humanities). In our analysis we aim to demonstrate an higher performance in those institutions specialized in scientific field. This variable is controlled as a dummy variable.

The last three control variables are looking at the knowledge orientation of University in different ways, but complementarily.

Scientific Sector Productivity

In the examination of potential influence of scientific sector, we took into account only scientific department (corresponding to Area 1-9 of the Italian system) and control for the assigned rating for each area. The rating has been provided by VTR2006[10] and it encloses several variables as for example the number of researchers, an evaluation of some research products, the international mobility of personnel, patents, spin-off and so on). Thanks to this synthetic scoring we look at the measure of a possible attractiveness of the university in terms of industrial funding. For this aim we have collected 1.236 departments, secondly we have clustered them into the 9 scientific area and assigned an average rating per area per University.

Table 1

Universities - Descriptive Statistics

Year	Frequency	Polytechnic	Non Polytechnic	Public	Private	Advanced School	Foreigners	Specialized Institutions	General studies
2006	69	4	65	58	6	3	2	15	54
2007	69	4	65	58	6	3	2	15	54
2008	69	4	65	58	6	3	2	15	54
2009	69	4	65	58	6	3	2	15	54
Total	**276**	**16**	**260**	**232**	**24**	**12**	**8**	**60**	**216**

As shown in table 1, sample composition of our data is balanced and all observations are equal per year.

Table 2

Variable	*Scale*	*Source*	*Measure*
DA	Continuous	NUCLEI MIUR	Degree of Autonomy
PROF	Discrete	PRO3	N° of Professor and researchers
EXT_PHD	Continuous	PRO3	N° of PhD Scholarship financed by external subjects
IF	Continuous	NUCLEI MIUR	Industrial funding
INT	Continuous	PRO3	International contracts and agreements
OTHER	Continuous	NUCLEI MIUR	Other income from
TER	Continuous	NUCLEI MIUR	Total Budget for Research
QUAL	Continuous	VTR CINECA	Quality index set by VTR

5 Empirical Analysis

In order to carry out the analysis, some new variables have been developed to mitigate the scale effects and to better represent structural university

[10] See also www.cnvsu.it

characteristics. We have adopted for estimation a regression model, supplemented by other estimations including control variables.

$$\frac{DA}{TER} = \alpha + \beta_1 \frac{PROF}{TER} \qquad (M1A)$$

$$\frac{IF}{TER} = \alpha + \beta_1 \frac{PROF}{TER} + \frac{INT}{TER} \qquad (M1B)$$

$$\frac{IF}{TER} = \alpha + \beta_1 \frac{GOV}{TER} + SIZE + NORTH + GENER + POLI \qquad (M2)$$

$$IF_{OTHERPROF} = \alpha + \beta_1 \frac{ECO}{PROF} + SIZE + NORTH + GENER + POLI \qquad (M3)$$

$$\frac{IF_{OTHER}}{TER} = \alpha + \beta_1 \frac{MIUR}{TER} + BTOT + SIZE + GEOG + TYPE + POLI \qquad (M4)$$

$$IF = \alpha + \beta_1 QUAL \qquad (M5)$$

From the observed results, we can note how total budget of research (TER) varies in function of size, staff and total expenses and in a general perspectives it grows in the selected period. Nevertheless there are consistent differences in the composition of Income for research activities. Descriptive shows a slight decline of private source in 2007, that significantly change its trends in the following years, while only Government funds decrease constantly over years.

Further, the distribution of all sources of funding is subjected to a great variability. Contracts with international private actors are constantly increased over the selected period, as well as resources from projects funded by European Union.

Table 3

YEARS/VALUE (K€)	ENTERP_	Delta	EU_	Delta	MINIST_	Delta	OTHER	Delta	Total ext_ resources	Delta
2006	52.375		118.637		244.187		363.128		769.636	
2007	48.977	-6%	158.192	33%	214.992	-12%	318.301	-12%	863.549	12%
2008	52.278	7%	111.711	-29%	161.340	-25%	320.730	1%	744.075	-14%
2009	57.180	9%	181.355	62%	155.585	-4%	418.366	30%	1.017.936	37%
AVERAG	52.703		142.474		194.026		355.131		848.799	
STAND_DEV_	3.377,1		33.032,1		42.824,4		46.913,0		123.906,9	

Funding distribution by year and source

Beyond the general funding distribution, if we consider indexes based on research activity input (as number of Professors and researchers, or number of PhD programs), it is of great interest to note that the variation in economic capacity per professor, inversely related to the size of the universities, but with significant exceptions that could be found in the top University table. This aspect could be associated to an higher efficiency of small structures, or alternatively, a structural barrier for structured behavior in the large universities. Literature does not report specific relationships between size (in terms of Number of staff) and efficiency, but limits its considerations to particular scientific sectors.

Table 4

		(Average values)		
Size	N° Prof	N° PhD Program	Expenses tot for research (k€)	Economic capacity per Prof.
Micro	177,1	8,5	4.490,6	32,4
Small	569,8	19,1	8.023,8	14,2
Medium	1.034,3	38,2	18.269,2	18,3
Macro	2.502,5	86,2	40.110,8	16,6

Funding resources and staff by size

Table 5

TOP FIVE	ENTERP_	TOP FIVE AS % OF TOT	TOP FIVE	OTHER INCOME	TOP FIVE AS % OF TOT
2006	**52.375**	**52%**	**2006**	**363.128**	**35%**
MILANO	13.618		Scuola Normale Superiore di PISA	39.983	
POLITECNICO TORINO	5.479		ROMA "La Sapienza"	31.631	
ROMA "Tor Vergata"	2.892		GENOVA	21.150	
FIRENZE	2.750		POLITECNICO MILANO	17.878	
BOLOGNA	2.631		BOLOGNA	17.129	
2007	**48.977**	**42%**	**2007**	**318.301**	**32%**
MILANO	7.840		ROMA "La Sapienza"	29.606	
POLITECNICO TORINO	4.282		POLITECNICO MILANO	22.469	
FIRENZE	3.445		GENOVA	18.237	
BOLOGNA	3.034		BOLOGNA	16.380	
POLITECNICO MILANO	1.876		PADOVA	15.779	
2008	**52.278**	**40%**	**2008**	**320.730**	**31%**
MILANO	8.635		POLITECNICO MILANO	22.470	
POLITECNICO TORINO	5.956		ROMA "La Sapienza"	21.005	
FIRENZE	2.530		GENOVA	19.486	
POLITECNICO MILANO	1.948		BOLOGNA	18.321	
VERONA	1.703		NAPOLI "Federico II"	17.991	
2009	**57.180**	**28%**	**2009**	**418.366**	**32%**
VERONA	5.226		POLITECNICO MILANO	34.638	
FIRENZE	3.788		ROMA "La Sapienza"	32.418	
POLITECNICO MILANO	2.807		BOLOGNA	25.415	
TORINO	2.196		NAPOLI "Federico II"	20.472	
SIENA	2.072		GENOVA	19.503	

Top Universities by private funds and other income (contracts, consultancy etc.)

Table 5 presents the best performing Universities per year and per source of funding. Top five universities represent a good percentage of the total in 2006 while in the following years private funding has definitively raised and more diluted over several universities. Another interesting (and largely investigated) facet is that almost all top Universities are from North Italy. According to literature this is coherent with other performances in third mission activities, and also in the case of private funds it is probably linked to entrepreneurial environment particular developed in regions as Lombardia, Emilia Romagna and Piedmont.

Consequently, we have controlled regression models for size and geography. In both columns it is straightforward that top universities do not change their positions across years.

In order to capture the average income deriving from industry we present a table with these values split by the more significant structural characteristic of Universities. In particular, in Table 6, we note a positive increasing of general institutions, versus an inverse trend in specialized institutions (all those Universities which limit their activity to a restricted range of topics).

Table 6

					Average Value (K€)					
Year	General studies	Specializ	Polytechnic	Public	Advanced Schools of Doct.	Private	Micro	Small	Medium	Mega
2006	807,7	673,6	1.910,3	883,8	69,7	151,2	151,3	356,3	859,8	2.458,9
2007	767,5	592,9	1.723,8	820,2	9,7	229,5	160,9	372,8	965,3	1.931,8
2008	795,2	723,9	2.179,0	876,9	19,7	226,0	204,3	445,3	1.097,9	1.812,0
2009	912,7	629,2	1.151,5	914,6	231,3	573,5	360,0	667,5	1.127,9	1.549,7

Average income from industry per University

As argued before, the trend suggests that leads University in exploiting also different area, while on the other hand, Polytechnics show a slight but constant decline in attracting these resources, while Private Universities and Advanced Schools of Doctorate represent a really interesting case to be further investigated.

As argued by Turri (2011) Advance Schools of Doctorate have interpreted better than others the "corporate model" described by Olsen (2007), partly due to their organization and their market driven attitude, there are able to attract several resources, and thanks to their small dimension and limited number of staff.

6 Results and Discussion

The results of the five models are presented in table 8. The estimation results include a Beta coefficient slight negative for the first two hypothesis and in case of Model 1b not significant. In fact, degree of autonomy and Industrial funds do not rely positively with professor efficiency and are not significant with international contracts.

Table 7

Variable	St. Dev	Mean	Confidence interval 95% Inf	Sup	Sig.
DA/TER	0,46	0,60	0,55	0,66	0,00
PROF/TER	0,05	0,06	0,06	0,07	0,00
IF/TER	0,06	0,05	0,04	0,05	0,00
GOV/TER	0,10	0,15	0,14	0,17	0,00
IF_OTHER/PROF	26,68	8,80	5,59	12,00	0,00
ECO/PROF	0,02	0,01	0,01	0,01	0,00
IF_OTHER/TER	0,43	0,32	0,26	0,37	0,00
PERF	0,05	0,03	0,03	0,04	0,00

Descriptive statistics

Table 8

Model estimation

	1	1B	2	3	4	5
PROF /TER	-,160 (0,546)***	-0,73(0,077)				
INT/TER		-0,19 (0,038)				
Gov/TER			-,122 (0,040)*		0,226 (0,269)***	
ECO/PROF				0,221(103,7)***		
BTOT					0,166 (0,500) ***	
VTR 2008						0,085 (64,7)**
VTR 2007						0,073 (35,07)**
VTR 2006						0,075 (41,23)**
Size 1			-0,123 (0,040)	0,195 (105,3)**	0,228 / 0,160 **	
Size 2			-0,124 (0,040)*	0,215 (104,8)**	0,226 / 0,168 **	
Size 3			-0,121 (0,040)	0,219 (104,6)**	0,227 / 0,164 **	
Size 4			-0,120 (0,040)	0,223 (103,8)**	0,227 / 0,164 **	
North			-0,123 (0,041)**	0,228 (104,7)**	0,234 / 0,165 **	
Generl			-0,118 (0,040)	0,178 (106,4)**	0,230 / 0,145 **	
Poli			-0,077 (0,040)**	0,220 (104,8) **	0,233 / 0,170 **	

****p-value < 0,01 (2-tails) **p-value < 0,05 (2-tails) *p-value < 0,1 (2-tails).*

Beta(std error)

Regression results

Secondly, Government fund distribution is negatively associated to Industrial funds, and the relations support our hypothesis of a fostering other external collaborations. We also observe a good deal with "North" dummy variable that do improve the model.

One of most significant results from the third model is the positive correlation between Economic availability of Professors and their effective attractiveness of industrial funds and other contracts. It seems that research oriented Universities are likely to perform better in acquiring new resources from private investors. We controlled the model for different variables and according to literature North dummy increase the robustness of the model. We also noted that mega University (size >40.000 students) and Polytechnic are slight sensitive to these collaborations. Switching to relation to overall performance including the structure employed for research, personnel dedicated and PhD scholarship we note an positive and associable relations with industrial funds. University with a better performance index tend to attract a huge amount of resource. The model is significant for all controlling variables.

We finally consider if quality evaluation is relevant with Industrial attractiveness and set the ranking score of scientific departments for all university included in the sample, we then examine relations between ranking score and industry funding to relative department. The result is significant and positive for all the three years considered (2006-08), while 2009 has been excluded for diluted time effects. Our conclusion is that industry funds are in a positive relation in the Models 3 4 and 5 and even though private source is a small component of entire budget, nevertheless it is affected by performance indexes and in particular ranking scores. These indexes can be considered as a proxy for internal organization of the University and are related to the external capacity of attractiveness.

On the other side, limiting the analysis to some sources of funds (Government, professor/economic resources) only we do not find a significant relation and in certain cases we have a slight negative results, probably due to the incompleteness of the built indexes.

7 Conclusions

From industry viewpoint, University represents a big resource of knowledge that can improve an existing product or capability (Bercovitz Feldman 2006). Many empirical researches can demonstrate it. Feller et al. (2002) reported that 63% of the companies participating at Engineering Research Centers (ERCs) have received direct technical assistance from university researches. Geuna (1999) and Mowery et. Al. (2001) underline that university has long served as a source of foundation scientific and technical knowledge; however the discovery of breakthroughs with significant commercial potential such as biotechnology, computer science, material science and nanotechnology is driving increased industry sponsorship of university research. The research type, university-industry interactions may differ in terms of level of ongoing involvement. A firm's R&D alliances with universities may involve either single transactions such as individual projects or in-depth long-term relationships as another part of R&D strategy (Berkovitz and Feldman 2006). In particular, among the several alliances that firm can carry on, universities are preferred partners when there are concerns about the perceived ability to fully appropriate the results, for those projects that engage a long-term and risky strategy. The difference between firm and university can provide a unique incentive, because the partner feature (public) can prevent the actor to act opportunistically, for example including the right of first refusal to license IP resulting from the project.

The aim of this paper is to analyze the evolution of Industrial source of funding and discuss the variables affecting the distribution of industrial funding with other University indicators, included quality indicators. We have found that only complex indexes that enclose several aspects of the research organization and performance could positively drive attractiveness of industrial funds. We also find that for scientific sector ranking scores are strictly related with industry attractiveness, while we exclude that single variables (as for example Number of staff or Government income) may generate a particular or positive influence. Our results show that a good University performance may facilitate an external collaborations, but definitively it can foster mutual benefit in University-Industry frame of relations.

Although the set of available data and the several Universities examined, this research presents some caveats. It tries to design a general situation, but at University level, marginally considering the department level, that generate a great variability across results. Further researches should be dedicated to the examination at a department level of different economic and organization attitudes. More attention should be devoted also to quality perception of scientific outputs by industry actors, by using not only economic variables but also multidimensional qualitative measures of Research activities, to identify the decisional process supported by industries in the choice of their academic partner.

References

1. Arundel, A., Geuna, A.: Proximity and the use of public science by innovative European firms. Economics of Innovation and New Technology, Taylor and Francis Journals 13(6), 559–580 (2004)
2. Arvanitis, S., et al.: University-industry knowledge and technology transfer in Switzerland: What university scientists think about co-operation with private enterprises (2008)
3. Baldini, N., Grimaldi, R., Sobrero, M.: Institutional changes and the commercialization of academic knowledge: A study of Italian universities' patenting activities between 1965 and 2002. Research Policy 35, 518–532 (2006)
4. Bercovitz, J., Feldman, M.: Fishing upstream: Firm innovation strategy and university research alliances. Research Policy 36, 930–948 (2007)
5. Bleiklie, I.: From Individual Pursuit to Organised Enterprise: Norwegian Higher Education Policy and the Postgraduate Curriculum. European Journal of Education 29(3) (1994)

6. Bonaccorsi, A.: Universities and strategic knowledge management. Elgar Publishing, UK (2007)
7. Bruneel, J., D'Este, P., Salter, A.: Investigating the factors that diminish the barriers to university-industry collaboration. Research Policy 39, 858–868 (2010)
8. Capano, G.: L'Università in Italia, il Mulino, Bologna (2000)
9. Capano, G., Regini, M.: Tra didattica e ricerca: quali assetti organizzativi per le università italiane? Le lezioni dell'analisi comparata, Roma, Fondazione CRUI (2011)
10. Caputo, A., Palumbo, M.: Manufacturing re-insourcing in the textile industry: A case study. Industrial Management & Data Systems 105(2), 193–207 (2005)
11. Cineca Website, `http://www.cineca.it`, `http://www.vtr2006.cineca.it` (accessed March 15, 2012)
12. Clark, B.R.: Diversification of Higher Education: Viability and Change. In: Meek, V.L. (1996)
13. Cohen, Levinthal: The absorptive Capacity: A new Perspective on Learning and Innovation. Administrative Science Quarterly 35, 128–152 (1990)
14. D'Este, P., Patel, P.: University-industry linkages in the UK: What are the factors underlying the variety of interactions with industry? Research Policy 36, 1295–1313 (2007)
15. de Boer, H.F., Enders, J., Leisyte, L.: Public sector reform in Dutch higher education: the organizational transformation of the University. Public Administration 85, 27–46 (2007)
16. Di Gregorio, D., Shane, S.: Why do some universities generate more start-ups than others? Research Policy 32(2), 209–227 (2003)
17. Etzkowitz, H.: An Entrepreneurial Science. Routledge (2003)
18. Etzkowitz, H., Leydesdorff, L.: Research Policy 29(2) (2000)
19. Feller, I.: New organizations, old cultures: Strategy and implementation of interdisciplinary programs. Research Evaluation 11, 109–116 (2002)
20. Fini, R., et al.: Complements or substitutes? The role of universities and local context in supporting the creation of academic spin-offs. Research Policy 40, 1113–1127 (2011)
21. Geuna, A.: The Economics of Knowledge Production: Funding and the Structure of University Research. Edward Elgar, Cheltenham (1999)
22. Geuna, A.: The Changing Rationale for European University Research Funding: Are there Negative Unintended Consequences. Journal of Economic Issues 35, 607–632 (2001)
23. Giuliani, E., Arza, V.: What drives the formation of 'valuable' university–industry linkages?: Insights from the wine industry. Research Policy 38(6), 906–921 (2009)
24. Grossman, Helpman: Income distribution, product quality, and international trade (1991)
25. Gulbrandsen, M., Slipersaeter, S.: The Third Mission and the Entrepreneurial University Model. In: Bonaccorsi, A., Daraio, C. (eds.) Universities and Strategic Knowledge Creation: Specialization and Performance in Europe. Edward Elgar Publishing, Cheltenham (2007)
26. Hauptman, A.M.: Higher Education Finance:Trends and Issues. In: James, J.F., Forest, J.J., Altbach, P.G. (eds.) International Handbook of Higher Education (2006)
27. Herbst, M.: Financing Public Universities: the case of performance funding. Springer (2007)

28. Lundvall, B.A.: National System of innovation-toward a theory of innovation and interactive learning. Pinter Publishers, London (1992)
29. Mansfield, E.: Academic research and industrial innovation: an update of empirical findings. Research Policy 26, 773–776 (1998)
30. MIUR. Italian Ministry of Instruction, University and Research, `http://www.miur.it` (accessed March 15, 2012)
31. Moscati, R., Regini, M., Rostan, M.: Torri d'avorio in frantumi? Dove vanno le università europee, Bologna (2010)
32. Mowery, D., Nelson, R., Sampat, B., Ziedonis, A.: Ivory Tower and Industrial Innovation: University-Industry Technology Transfer Before and After the Bayh-Dole Act in the United States. Stanford University Press, Stanford (2004)
33. OECD, Frascati Manual. Proposed Standard Practice for Surveys on Research and Experimental Development. OECD, Paris (2002)
34. Olsen, J.: European Debates on the Knowledge Institution: The Modernization of the University at the European Level. Higher Education Dynamics 19(Pt. 1), 3–22 (2007)
35. Palumbo, R.: L'Università nella sua dimensione economico aziendale. Evoluzione sistemica e modelle razionalizzanti, Giappichelli (1999)
36. Salmi, Hautman: Innovations in Tertiary Education Financing: A Comparative Evaluation of Allocation Mechanisms. Working paper series world Banck (2006)
37. Slaughter, S., Leslie, L.: Academic Capitalism. Johns Hopkins University Press, Baltimore (1997)
38. Turri, M.: L'università in transizione. Guerini Studio (2011)
39. Zucker, L.G., Darby, M.R., Armstrong, J.S.: Commercializing knowledge: University science, knowledge capture, and firm performance in biotechnology. Management Science 48(1), 138–153 (2002)

Breaking Down Barriers and Building Collaboration

Knowledge Transfer Success Stories from Wales

Jarmila Davies

Academia Engagement Senior Manager
Department of Business, Economy, Technology & Science
Welsh Government

Abstract. The Welsh Government is fully committed to supporting development of knowledge economy and encouraging academic institutions to play a greater role in economic and social development of Wales by aggregating resources, reducing duplication of initiatives and encouraging creative expansion of innovation in Welsh businesses and academia.

Development and creation of the knowledge transfer continuum is a long term task that requires not only dedication and passion of all involved but also necessary funding to start the process going.

This paper highlights commitment and support the Welsh Government is giving to this highly desirable course of action.

1 Background

The Welsh Government (WG) [1] has been fully committed to supporting the development of a knowledge economy in Wales for over a decade.

The initial engagement with the EU Regional Technology plan in 1997 laid the foundations for subsequently more complex and ground-breaking knowledge-exchange activities. Back then, Wales was selected as one of the economic regions in Europe with a remit to explore the extent of business / academia collaboration, and establish a knowledge transfer (KT) infrastructure.

The conclusion of the initial review indicated a patchy & uncoordinated level of activity and support infrastructure. Welsh businesses called overwhelmingly for a dedicated brokerage service, including a central point of information on academic resources available for commercial exploitation.

2 The Know-How Wales Programme

The Welsh Government adopted these recommendations, and established a dedicated KT support service, with a budget of £ 350K pa. The Know-How Wales (KHW) program was launched in 1999 and ran until 2008. A team of 6 dedicated KT specialists was located across the key economic regions of Wales.

R.J. Howlett et al. (Eds.): *Innovation through Knowledge Transfer 2012*, SIST 18, pp. 227–232.
DOI: 10.1007/978-3-642-34219-6_25

The principal aims of the programme were to;

- act as an impartial service and catalyst for change between academia & business
- Improve the coherence of service provision and focus on business needs
- facilitate access to commercially available academic resources
- manage and disseminate information on KT provision in Wales

The main activities were;

- Facilitating collaborative projects, strategic relationships, access to academic resources, and collaborating with the existing networks and fora
- Holding business breakfast meetings, problem-solving surgeries and bridging sessions
- Developing an information portal as the central point of access for knowledge and commercial academic recourses
- Establishing a dialogue with the FE sector in Wales

The outputs were encouraging;

- 796 collaborative R&D projects facilitated
- 1,770 businesses and academic institutions assisted
- 5,344 technical enquiries processed
- 208 jobs created.

In sum, the KHW project played a significant role in establishing a Knowledge Transfer continuum, and laid foundations for the development of a more comprehensive KT support service in Wales.

3 The Knowledge Exploitation Fund

The Knowledge Exploitation Fund (KEF) was established in 2000 by the Welsh Government, with the aim of enabling Higher Education (HE) and Further Education (FE) institutions to fulfill the prominent role expected of them in its economic development agenda. The KEF ran until 2007, with a combined total investment from the Welsh Government and the ERDF of £ 53.7m.

The principal aims of the KEF were to

- Encourage academia to create a culture of entrepreneurship and innovation
- Facilitate the commercialization of knowledge
- Transfer knowledge and expertise to industry
- Develop the business skills of academic staff and students

KEF operated alongside the Higher Education Funding Council for Wales's (HEFCW) Third Mission Fund. KEF provided pump-priming funds, whilst HEFCW supplied the core and mainstream funding.

There were several funding strands within the KEF offering:

3.1 *Culture Change and Capacity Building*

This strand helped senior management in academia develop and implement strategies for innovation and entrepreneurship. Creating *Enterprise clubs* within academic institutions enabled access to specialist support for business start-ups, fostered business links with local SMEs, and gave access to role models.

Entrepreneurial Champions were located within academic institutions to improve awareness and appreciation of entrepreneurship and innovation amongst staff and students.

3.2 *Commercialisation and Knowledge Transfer*

The aim here was to accelerate the commercialisation of ideas and products, and to cultivate a Knowledge Transfer continuum in Wales. A *Entrepreneurial Scholarship Scheme* offered budding entrepreneurs an allowance up to £6.5K to support new business start-ups.

Centers of Expertise and Knowledge Transfer provided incubation facilities to support fledgling knowledge-based enterprises. Located within academic institutions, they helped develop innovation and entrepreneurial skills.

Collaborative Industrial Research Projects (CIRP) were intoduced. This concept supported collaborative development projects between academia and a number of companies, including SMEs, as part of a single supply chain or market segment. CIRPs demonstrated that a specific academic problem was more likely to lead to commercial exploitation when businessses were involved at the outset. Every £1m invested in CIRPs resulted in £1.24m of private sector leverage, and the creation of 27 jobs and 20 jobs safeguarded [2].

The *Patent and Proof of Concept (PPoC)* was the first initiative of its kind in Wales. Its aim was to assist Welsh academic institutions develop IPR exploitation in-house, and to exploit academic research to maximum effect through specialist support.

Because the commercial exploitation process takes a long time, it is still too early to conduct a meaningful cost-benefit analysis of PPoC. To date, however, 17 patents or trademarks have been protected, with licensing deals and spin-out companies creating new opportunities for commercialisation [2]. Their impact should be monitored over the longer term.

The development of *Technology Transfer Networks (TTN)* facilitated joint working between research and technology transfer centers, providing an integrated and Wales-wide service to businesses.

3.3 *Engagement of the Further Education Sector in KT Activities*

It had been felt that the Welsh Further Education (FE) sector did not have a sufficiently robust supporting infrastructure for the delivery of collaborative academic projects, although they had the enthusiasm and will to develop these.

The Welsh Government was keen for the FE sector to engage in knowledge transfer & exchange activities. To this end, a *Knowledge Transfer Partnership (KTP) Mentoring project* was created, with a budget of £140K.

The principal aims of the *KTP Mentoring project* were to

- Engage the FE sector in KTP delivery
- Encourage a spirit of collaboration between HEIs and FEIs
- Develop FEIs staff confidence in initiating and running collaborative projects.

The Welsh Government provided overall project management and guidance, with leading HEIs being funded to deliver the project. KTP Advisers and Welsh Government specialists were available for specific support, such as application procedures and engagement with companies. The project also developed and delivered bespoke training for FE staff in collaboration with the KTP central office.

As a result, seven FE institutions out of 16 are actively involved in the KTP program and several more have included KTP delivery within their strategic plans. The UK Strategic Review of the KTP [3] and also the recent Review of the KTP delivery in Wales [4] showed that the FE sector in Wales is outperforming all regions in the UK, with the Gower College, Swansea being a leading KTP deliverer.

Clearly, a targeted support programme for a specific sphere within the KT continuum can strengthen the effectiveness of the whole.

4 The Academic Expertise for Business (A4B) Program

Reviews of both the KHW and KEF programmes indicated that although good progress has been made, additional funding was required to embed and extend already well-developed KT processes.

Building on experience, lessons learned and feedback from both businesses and academia the WG developed a new ERDF-funded programme called Academic Expertise for Business (A4B). It was launched in 2007 and will run until 2014, with a budget of £70m.

There are two principal strands of the programme, each aiming to further develop and strengthen the KT continuum in Wales.

4.1 Knowledge Exploitation Capacity Development

The ability to develop a sustainable knowledge economy in Wales depends on the capacity of KT practitioners to understand -and effectively manage - collaborative activities with businesses.

The key objectives of this strand are to help academic institutions embed and strengthen their capacity for exploiting knowledge alongside companies, strengthening R&D, innovation and commercial exploitation in the Convergence Area.

A4B funding was used to provide grant support to academic institutions for KT practitioners to undertake professional development training, aimed at increasing their skills and knowledge of innovation technology management.

Based on the AURIL CPD framework and working with the Institute of Knowledge Transfer (IKT), the WG provided initial funding to Cardiff University to develop a bespoke foundation course for KT practitioners. The course enabled each attendee to develop an individual CPD plan which built on their professional strengths and future professional aspirations.

KT practitioners were encouraged to select an IKT-approved relevant course to underpin their future development. 45 KT practitioners from the HE and FE sectors completed the course and gained IKT membership.

Feedback from the course indicated an awareness that a knowledge-based society operates on a different dynamic to an industrial society focused on the production of tangible goods.

The course has been accredited by the IKT and is now available nationally to anybody who wishes to progress a career in KT.

4.2 Knowledge Transfer and Collaborative Industrial Research

The principal aims of this strand are to enhance and accelerate knowledge & technology transfer from Higher or Further education institutions to key sector businesses in the Convergence Area. Half way through the six-year programme, there have been over 400 projects supported.

The Welsh Government has been engaging with the Technology Strategy Board in the development and delivery of the shorter KTP programme, available between 2009 and 2011. This has been very popular with Welsh companies because it was tailored to their specific needs and level of business development. The recent strategic review of the shorter KTP programme indicated that every £1 of grant invested by the WG generated £37.60 of additional business turnover and £9.43 of additional GVA, with 131 additional jobs created [5].

Finally, an enabling tool across the continuum has been the development of the knowledge information portal, Expertise Wales; www.expertisewales.com. The portal offers a central point of access to expertise and facilities available in universities and colleges in Wales, with key contact information. The content of the portal, including information, tools and links, is provided free of charge to all parties seeking involvement in knowledge transfer activity.

5 Conclusion

Research and Development (R&D) plays an important role in stimulating enterprise and innovation. Innovation is a key driver of productivity, economic growth and long term improvements in well-being. Wales is moving towards a more R&D-intensive and knowledge-based economy where innovation can flourish.

The Welsh Government is fully committed to supporting these developments by encouraging academic institutions to play a greater role in the economic and social development of Wales by aggregating resources, reducing duplication, and encouraging creative interaction between Welsh business and academia.

References

[1] Economic Renewal: a new direction. Welsh Government (July 2010)
[2] Review of the Impacts of the ERDF Objective 1 Projects in Phase 2 of KEF. Jenesys Associates Ltd. (August 2007)
[3] Knowledge Transfer Partnership Strategic Review. Regeneris Consultng (February 2010)
[4] Evaluation of the Classic Knowledge Transfer Partnership in Wales. CM International UK Ltd. (December 2011)
[5] Evaluation of the Shorter Knowledge Transfer Partnership in Wales. CM International UK Ltd. (December 2011)

SDL Approach to University-Small Business Learning: Mapping the Learning Journey

Christopher J. Brown and Diane Morrad

University of Hertfordshire, de Havilland Campus, Hatfield, Herts, AL10 9AB

Abstract. This paper explores the important link between the knowledge exchange activities of small businesses and universities, and the co-production and co-creation of value as perceived by the small business owner-manager. Small business owner-managers seek out information to help identify opportunities, especially information that formalizes their mental schemas around positive outputs and outcomes associated with innovation adoption and dissemination. More importantly, these same owner-managers identify advisors within these knowledge exchange encounters that will help them develop their mental schemas to understand the requirements for change, analyse the existing or latent market needs, and through this develop new understanding. We present our findings in the form of a map detailing the co-production/co-creation of value derived from university-small business collaboration, and some insights into the motivation, rationale and experiences of both parties. We conclude with our understanding of the outcomes and impacts derived, and suggestions on how the collaborative partners could better manage the whole process.

1 Introduction

The UK economy is on the edge of an important tipping point, one where large and small enterprises alike are balancing the "needs of a firm's direct and indirect stakeholders (such as shareholders, employees, clients, pressure groups, communities, etc.), without compromising its ability to meet the needs of future stakeholders as well" (Dyllick and Hockerts 2002). In practical terms this means addressing the 'triple bottom line', where economic, social and environmental needs are balanced – financial benefits/costs against social/environmental benefits. This is very difficult for SMEs with limited resources, and pressure from their supplier/customer to focus more on cutting-costs and prices (Revell and Blackburn 2007). General research carried out by the government and other interested parties has highlighted the broad issues of small business awareness and engagement with these eco-innovation initiatives (DEFRA 2006; BERR 2010). Other sector-based research has identified one major factor contributing to the slow transitioning of the SME sector: awareness and lack of information associated with creating and developing a business case for eco-innovations(Adams, Hammond et al. 2011). In part this is probably down to SME owner-managers existing schemas, their organization of information in a given area, for example eco-innovations.

R.J. Howlett et al. (Eds.): *Innovation through Knowledge Transfer 2012*, SIST 18, pp. 233–243.
DOI: 10.1007/978-3-642-34219-6_26

Those owner-managers who already are alert to the need to engage in eco-innovations are already seeking out information that both challenges their existing mental schemas, and identifies more complex information, and where they should get this, that will stimulate an objective assessment of the 'nature of the change needed' (Ozgen and Baron 2007). SME networking activities are critical to the acquiring of this new knowledge (some of which is market related), additional resources and importantly mentoring opportunities; knowledge exchange schemes operated by business-facing universities can fulfil this role (Carson, Gilmore et al. 2004). In the broad topic of SME networking we are exploring one element - that of small businesses and universities, studying two collaborations over an 18-month period.

2 Small Business Networking and Aggregation

Since the 1980's Networks have become an organizational form that is increasingly important where businesses are putting together complex products requiring a myriad of different bought-in items. The locus of innovation and adoption for these market leaders and their competitive products/services are these networks they either create or engage with (Lockett and Brown 2006). Small businesses engagement with external resources is significantly influenced by their level of development, both what is happening internally and externally (Dyer and Ross 2007). Small businesses exhibit different problems associated with start-up than later in their development, and as such this will significantly impact on the small business owner-advisor relationship. Some small business research suggests that age and size of firm is a factor in their choice of seeking and working with external advisors (Jay and Schaper 2003). They suggest that small businesses that are growing, and owner-managers that are confident, are more likely to regularly accept external advisors and use them for shorter durations than other business owners.

Too little research exists on these small business owner-advisor relationships specifically associated with university-business collaborative projects (Dyer and Ross 2007). Two important variables are key to the effective co-creation of value in these collaborative projects: the first is the time it takes for small business owners to generally accept and trust the other advisor/partner; the second is the durability of that relationship, does it last over the entire length of the project and beyond?

Similar to larger organizations, small businesses use aggregations to help reduce costs, probably increase revenue and often mitigate risks around economic and environmental responsibilities (Lockett and Brown 2006).

SME networking can be further classified based on the degree of formality and integration, certainly this identifies and evaluates the level of commitment and engagement of the different aggregations (Brown and Lockett 2004).

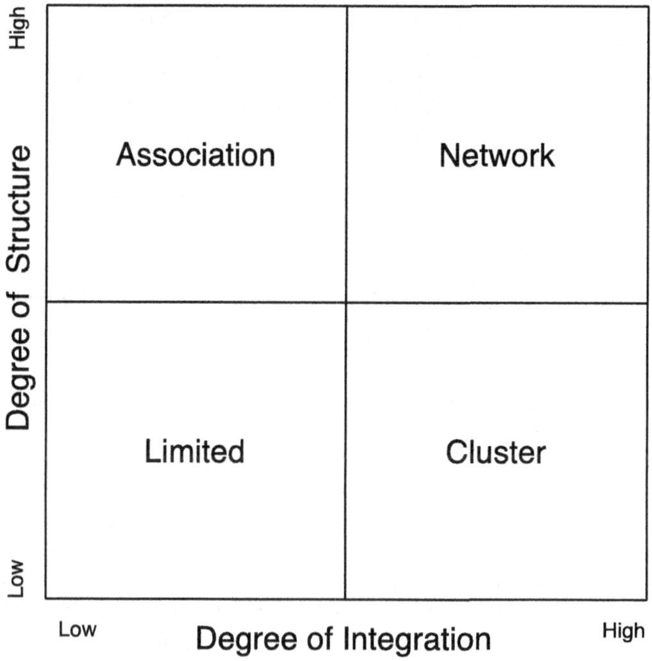

High
Degree of Structure
Low
Association
Network
Limited
Cluster
Low
Degree of Integration
High

Fig. 1 Taxonomy of Aggregations for SMEs

These different types of aggregation help distinguish the level of engagement and role of the advisor in supporting and advising these businesses, and also what the business expectations are regarding favourable outcomes:

Limited any SME relationship that is both informal and independent of any full commitment;

Association more formal relationships requiring annual membership or small financial commitment, again interactions are non-committal;

Cluster more formal commitment to the relationship and a degree of dependence is built up between the partners and members of the aggregation;

Network more structured and dependency-based, often involving formal and legal collaborative projects – working together around some shared vision and value (joint enterprise).

Significantly, the role of the external party in any relationship with the small business owner-manager is an exchange of services, these services are in terms of eco-innovation initiatives focused around knowledge and expertise. Like in the example of small businesses networking, it is often to gain greater understanding of the outcomes and impact of moving towards a e-business model, small business owner-managers evaluating these eco-innovation initiatives are often seeking three types of service, not necessarily from just the one party:

Knowledge a party who can provide knowledge and expertise on adopting and diffusing eco-innovations;

Enterprise a party who can actually provide the resources to adopt and implement the eco-innovative technologies and processes;

Community a party who can help connect the business owner-manager to other businesses who can either help with co-creating value in the product or service offering, or offer additional knowledge and expertise that will reduce the overall risk of the innovation's adoption and diffusion.

In the next section, the value of university-small business collaborative schemes are particularly explored.

3 Knowledge Exchange – Universities as Part of SME Networks

Knowledge Exchange (KE) schemes within universities are not new. Traditional universities used these types of schemes to both channel their intellectual property, and look at improving the business impact of their research outputs through related consultancy work. Knowledge exchange across the business community is based on a phenomenological paradigm, one that includes a post-realism perspective of what is real and realistic in practice (Vargo and Lusch 2008). SME owner-managers use mental frameworks, schemas, to organize the information they retain, about marketplaces, opportunities and threats (Wyer and Srull 1994). These owner-manager mental schemas help structure existing information and identify gaps or challenges. Those owner-managers who are more alert to the need to respond to eco-innovation challenges positively seek out new information to help them understand the potential adjustments needed to their existing schema (Ozgen and Baron 2007). It is these SME owner-managers that often make the first approach to engage in knowledge exchange projects. They are seeking more complex information to both objectively assessed the nature of change, and importantly understand the potential outcomes and impact it will have on their industry and the business(Gaglio and Katz 2001).

Accessing knowledge and expertise is important for SMEs faced with the challenge of integrating sustainability into their business practices (Jamsa, Tahtinen et al. 2011). Research on SMEs relating to the value of the networking, demonstrates the relationship between identification of opportunities and access to information (Fuller and Tian 2006; Ozgen and Baron 2007). SME networks create valuable business relationships through which they gain valuable information, skills and competencies. SME networks are thus used to (Chaston and Mangles 2000; Gilmore, Carson et al. 2001):

- Acquire market knowledge;
- Other information needed to aid business decision-making – critical to opportunity recognition (Ozgen and Baron 2007);
- Obtain valuable resources;
- To help specialize;
- To increase cost-efficiency;
- To learn from others.

The SME owner-managers perception of the external world and even knowledge of themselves, as perceived by others and furnished to them through information exchange during social networking, is most valuable (Ozgen and Baron 2007). Many of these SME owner-managers turn to older and more experienced individuals who assist them in the task of acquiring useful skills and knowledge (Clutterbuck and Ragins 2002).

In the Small- to Medium-sized Enterprise sector, this is especially so, their experiential learning points them towards relationships that create and develop value. This is especially true where interactional processes associated with their respective service systems, help the parties become co-creators of value. This is effectively why so many small business owner managers find the relationship and value derived from university-small business collaborations so valuable and beneficial.

4 New Service Dominant Logic – Focused on Co-production/Co-creation of Value

In discussing the knowledge exchange partnerships in the above section, it was noted that the knowledge and expertise being exchanged involved two service systems, that of the universities, and those of the small business. In the Service Dominant Logic (S-D Logic) literature, it has been suggested that the journey's undertaken as part of this exchange need to be mapped alongside the types of encounter involved and ultimately the value exchanged (Payne, Storbacka et al. 2008).

The authors wanted to be able to define the basis of the value proposition and the value-in-use exchanged in these university-business collaborative projects (Lusch & Webster, 2011, Frow & Payne 2008, 2010) in order to achieve congruence between stakeholder perceptions, image and operational activities (Lomax and Mador, 2006; Lusch and Webster, 2011) and thus be able to define the processes needed to support co-creation.

A fundamental premise of SD-logic is that the definition of value changes continuously (Lusch & Webster, 2011). How do the actors (owner-managers and advisors) involved evaluate what they are doing (Naidoo, 2010) and how can understanding this part of the process enable knowledge exchange activities to create sustainable value over time? The concept of a continuum, with co-production at one end and co-creation at the other, reflects the possible degree of relationship between a university and small business (Cova and Salc 2008).

Table 1 Taxonomy of Co-Production/Co-creation in Knowledge Exchange

Attributes/Measures	Co-Production	Co-Creation
Knowledge	Existing	New
Innovation	Incremental	Radical
Business Model	Adapted	Change
Impact	Local/short	Boundary spanning/long

5 Research Methodology

SD-Logic is attractive when considering the implications of knowledge exchange projects, which by definition require co-creation (Vargo and Lusch, 2004; 2010) in order to succeed. The conceptual nature of the S-D logic theory presents opportunities for further analysis regarding how the ideas apply in a variety of real world settings: understanding the nature of the relationships required to support effective co-creation could enable better processes to be developed to support Knowledge Exchange activities. The aim of the research was therefore to test the concept of service dominant logic (S-D logic) as a basis for university business Knowledge Exchanges using empirical data sourced from Knowledge for Business (K4B) activities.

5.1 Objectives

1. To understand the process of co-creation/co-production in specific knowledge exchange activities
2. To determine the roles of the different actors in the co-creation/co-production process
3. To analyse how the co-creation process can be managed and supported more effectively

5.2 Approach

An ethnographic and grounded theory approach (Pettigrew 2000) was adopted in order to develop thick data from individual projects that could then be used as a basis for evaluating the phenomenon of S-D logic in knowledge exchange activities. Both methods are considered suitable for explicating relationships (Huberman and Miles 1994). Case study research offers flexibility through the opportunity to combine data collection techniques (Darke, Shanks et al. 1998) which was attractive to the researchers when dealing with organisational contexts that varied considerably. It is the most popular social science research approach used to appropriately, and adequately, answer research questions associated with the interaction of eco-innovation initiatives and the businesses' underlying processes and systems (Darke, Shanks et al. 1998)

The aim was to test SD logic - a conceptual theory - in specific circumstances, producing empirical evidence to evaluate the theory. This combined ethnographic - grounded theory approach seemed particularly appropriate given its' ability to provide evidence "in areas where existing knowledge is limited" (Darke, Shanks et al. 1998) thus supporting a "roadmap" for theory development (Eisenhardt, 1989).

5.3 Limitations

The main constraints of a combined ethnographic and grounded theory approach relate to the extent to which the data can be considered generalizable

(Johnson 1990; Glaser 1992; Goulding 1998). In order to attempt to minimise this limitation the research was restricted to one aspect of the taxonomy of co-creation/co-production of value: the focus was on the understanding and interpretation of value as part of measuring outputs, outcomes and overall impact. The paper focuses on two completed case studies but the research is designed to be on-going with future K4B projects in order to develop the data set as a basis for better understanding of the construct of value within the co-production/co-creation process.

5.4 Instrumentation

In order to gain the necessary in depth understanding of the phenomenon of value as part of the co-creation process (Cavaye 1996) a combination of data collection techniques were used including interviews, observation, document and text analysis. This is in line with the researchers having limited prior knowledge of the constructs and variables which may be relevant (Benbasat, Goldstein et al. 1987; Cavaye 1996) and allowed for approaches relevant to the organisations being reviewed to be developed.

6 Key Findings

The longitudinal studies conducted were with two companies in the printing and social housing sector that had already engaged in previous knowledge exchange projects with the university, thus providing the authors with further insights into their value expectations from a further collaborative project.

Table 2 Summative Map of the Co-creation/Co-production of value for Universities and Small Business (Morrad and Brown Continuum)

Case Studies Outputs/ Outcomes and Impacts	Co-production	Co-creation
Knowledge	**DPG** Constrained thinking regarding opening up to new information sources or experiences outside of their own	**PH** Traditionally been strong on developing appropriate networks to leverage information, first time coming to a university. Positive about the benefits with working across schools on all aspects of the businesses growth challenges.
Innovation	**DPG** Restricted by their willingness to change organizational structures and operational processes in order to optimize the opportunities presented by new technology;	**PH** Innovation is often difficult in a tightly controlled market, where regulations and welfare acts closely control design. Yet, they are open to new materials and processes than deliver better performance, and at the same time competitive advantage.

Table 2 (*continued*)

Business Model	**DPG** Initial agreement on adopting the principle of market orientation but without understanding the need for cross-functional implementation – essential if they were to achieve full customer orientation;	**PH** Hard work developing a modular system that works for most applications – what they have found difficult is how to turn this into a sustainable business model. Yet openness to change and different ways to create new markets here in the UK and overseas has strengthened their BM.
Impact	**DPG** The initial market audit led to a better understanding of the market environment and opportunities and threats to the business. However, this led only to a short-term gain as there was no follow through in terms of the required organizational change to leverage opportunities and minimize threats.	**PH** The enterprise feels more confident in its self and its workforce and is relocating in more attractive premises to facilitate 100% growth over the next 3 years.

7 Discussion and Conclusions

This paper has presented evidence of the very different attitudes and practices of two small businesses, ultimately faced with the same challenge of evaluating and adopting eco-innovation measures – e-business and green products. Our two owner-managers showed a very different perspective to the willingness to open up the legitimate business case for changing their product/service offering, this is a common finding/conclusion from other research conducted across various sectors (Revell and Blackburn 2007).

These same studies point out the existing government policy of assuming that if small businesses are presented with a 'win-win' proposition then they will adopt these eco-innovations. With this in mind the government has adopted a strategy of providing more and more information, and additional incentives through 'FiT' and other favourable tax incentives. Yet, small business behaviour has not massively changed, and some of the reason is that the owner-managers mind-set schemas have not changed. More help is needed for businesses to understand how their business models will change, for some sectors and businesses this will initially lead to a decrease in margins, yet there are opportunities here too. This is were the differences lie between those businesses that look externally but close their minds to real change – co-producers; and those that are really engaged in opportunity recognition and are co-creators?

7.1 Co-production – The Typical Outcome from Knowledge Exchange?

In many of the small businesses that have developed over time, creating strong niche market position resulting in little need for extensive external links, any external links they have are both informal and require little or no commitment:

- Most of the knowledge emerging from the collaboration project is existing in that it is tacit within the organization, but the value of making it explicit effectively forces it into the open and demands that they face up to the challenge. Another benefit of this knowledge collecting process is that it forces a recognition of what is fact and what is myth. Importantly collecting this knowledge together, when often it sits within the different functional silo's;
- However, these business owner-managers are often in capable of using this knowledge quickly and effectively, unwilling often to disseminate it down to the relevant people and so fail to exploit its full potential.

Co-production benefits and costs:

Many of the small business owner-managers who are struggling with day-to-day fire-fighting activities and a business model that is consistently being challenged see:

- co-production as superficial and does not move the business on;
- representing a short-term limited adaptation to immediate challenges and priorities, but without fundamentally changing the ability of the business to respond to future threats/opportunities;

7.2 Co-creation – The Ideal Outcome from Knowledge Exchange?

Small business owner-managers who have an adaptable mind-set towards change, and focused on opportunity recognition:

- are open and able to assimilate and use new information or to respond to new perspectives on existing information;
- equally are keen to quickly develop, disseminate and exploit the knowledge to innovate faster than their competitors.

Co-creation benefits and costs:

Ultimately these owner-managers are also more likely to see that:

- Co-creation results in a more substantial change in overall perception of market needs, significantly changing both strategies and operations to deliver these;
- The process actually creates/forces a generative learning approach to forthcoming opportunities and threats;

Previous studies on university-business collaboration have highlighted selective factors that effectively create barriers to a fully successful collaboration, at least meeting the initial expectations of both parties. Our findings and conclusions suggest that small business networking, where that network includes universities has the potential to dramatically increase the opportunity for change, but only if these small business owner-managers can surmount:

1. That important external knowledge is gained by the small business, and this information can be and often is used by the business to create new knowledge to help create new novel products and service combinations, combining existing product/service attributes with new external elements. These co-creating owner-managers are usually time-sensitive and aware of the importance of innovating faster than their competition, a pointed noted in entrepreneurship research (Aldrich and Martinez 2001). The result of this innovation stemming from the SME's extended network, now including the university, is a set of new social connections, importantly linking people and resources (Obstfeld 2005).
2. That differences between business owner-managers' approaches to accessing and adoption of knowledge reflects their perception of likely business outcomes and impacts. In the one business they perceived it as incremental innovation, taking ideas and turning them into products using their existing processes and overall business systems. In the other business this new knowledge triggered an entire iterative process of re-evaluating the marketplace as showing opportunities for new products/services, hence embracing the need for new business models.

References

Adams, P.W., Hammond, G.P., et al.: Barriers to and drivers for UK bioenergy development. Renewable and Sustainable Energy Review 15, 1217–1227 (2011)

Aldrich, H.E., Martinez, M.A.: Many are Called, but Few are Chosen: An Evolutionary Perspective for the Study of Entrepreneurship. Entrepreneurship: Theory & Practice 25(4), 41 (2001)

Benbasat, I., Goldstein, D.K., et al.: The Case Research Strategy in Studies of Information Systems. MIS Quarterl 11, 369–386 (1987)

BERR. SMEs in a Low Carbon Economy (2010), `http://www.berr.gov.uk/files/file49761.doc` (retrieved February 04, 2010)

Brown, D.H., Lockett, N.: Potential of Critical e-applications for Engaging SMEs in e-business: a provider perspective. European Journal of Information System 13(1), 21–34 (2004)

Carson, D., Gilmore, A., et al.: SME marketing networking: a strategic approach. Strategic Change 13(7), 369–382 (2004)

Cavaye, A.L.M.: Case Study Research: A multi-faceted research approach for IS. Information Systems Journal 6, 227–242 (1996)

Chaston, I., Mangles, T.: Business Networks: assisting knowledge management and competence acquisition within UK manufacturing firms. Journal of Small Business and Enterprise Development 7(2), 160–170 (2000)

Clutterbuck, D., Ragins, B.R.: Mentoring and diversity: an international perspective. Butterworth-Heinemann, Oxford (2002)

Cova, B., Sale, R.: Marketing Solutions in Accordance with the S-D logic: Co-creating value in customer network actors. Industrial Marketing Management 37, 270–277 (2008)

Darke, P., Shanks, G., et al.: Successfully Completing Case Study Research: Combing, Rigour, Relevance and Pragmatism. Ino Systems Journal 8, 273–289 (1998)

Darke, P., Shanks, G., et al.: Successfully completing case study research: combining rigour, relevance and pragmatism. Information Systems Journal 8(4), 273–289 (1998)

DEFRA, Encouraging Sustainability amongst Small Businesses (2006), `http://randd.defra.gov.uk/Document.aspx?Document=SD14007_3807_INF.pdf` (retrieved January 12, 2011)

Dyer, L.M., Ross, C.A.: Advising the Small Business Client. International Small Business Journal 25(2), 130–151 (2007)

Dyllick, T., Hockerts, K.: Beyond the Business case fo Corporate Sustainability. Business Strategy & the Environment 11, 130–141 (2002)

Fuller, T., Tian, Y.: Social and Symbolic Capital and Responsible Entrepreneurship: An Empirical Investigation of SME Narratives. Journal of Business Ethics 67(3), 287–304 (2006)

Gaglio, C., Katz, J.: The psychological Basis of Opportunity Identification: entrepreneurial alertness. Small Business Economics 16, 95–111 (2001)

Gilmore, A., Carson, D., et al.: SME marketing in practice. Marketing Intelligence & Planning 19(1), 6–11 (2001)

Glaser, D.A.: Basics of Grounded Theory Analysis. Sociology Press, Milley Valley (1992)

Goulding, C.: Grounded Theory: The missing methodology on interpretivist agenda. Qualitative Market Research: An International Journal 1(1), 273–289 (1998)

Huberman, A.M., Miles, M.B.: Data Management and Analysis Methods. In: Denzin, N.K., Lincoln, Y.S. (eds.) Handbook of Qualitative Research, Sage Publishers, Thousand Oaks (1994)

Jamsa, P., Tahtinen, J., et al.: Sustainable SMEs Network Utilization: the case of food enterprises. Journal of Small Business and Enterprise Development 18(1), 141–156 (2011)

Jay, L., Schaper, M.: Which Advisers do Micro-firms use? Some Australian Evidence. Journal of Small Business and Enterprise Development 10(2), 136–143 (2003)

Johnson, J.C.: Selecting Ethnographic Informants. Sage Publishers, California (1990)

Lockett, N., Brown, D.H.: Aggregation and the Role of Trusted Parties in SME E-Business Engagement. International Small Business Journal 24(4), 379–404 (2006)

Obstfeld, D.: Social Networks, the teritus iungens orientation, and involvement in innovation. Administrative Science Quarterly 50, 100–130 (2005)

Ozgen, E., Baron, R.A.: Social Sources of Information in Opportunity Recognition: Effects of mentors, industry networks, and professional forums. Journal of Business Venturing 22, 174–192 (2007)

Payne, A.F., Storbacka, K., et al.: Managing the Co-creation of Value. Journal of the Academy of Marketing Science 36, 83–96 (2008)

Pettigrew, S.F.: Ethnography and Grounded Theory: A Happy marriage? Advances in Consumer Research 27, 1–10 (2000)

Revell, A., Blackburn, R.: The Business case for Sustainability? An examination of small firms in the UK's Construction and Restaurant Sectors. Business Strategy & the Environment 16(6), 404–420 (2007)

Vargo, S.L., Lusch, R.F.: Service-dominant logic: continuing the evolution. Journal of the Academy of Marketing Science 36, 1–10 (2008)

Wyer, R.S., Srull, T.K.: Handbook of social cognition. Hove, Lawrence Erlbaum Associates, Hillsdale, N.J (1994)

Innovative Challenges in Indian University Education: Agenda for Immediate Future

M.L. Ranga

Guru Jambheshwar University of Science and Technology, Hisar-125001 (Haryana) India

Abstract. The spoken agreement between society and university education that provided expanding resources in return for greater right of entry for students as well as researchers and service to society has gone down, which at this moment demands some retrospection before setting for future agenda in this regard. This paper discusses some of the innovative challenges of university education in country like India. Challenges such as autonomy and accountability, the role of research and teaching, academic reform and the curriculum, and the implications of the massive expansion that have characterized universities in many countries are of primary concern here.

Keywords: Innovative challenges, Indian universities, Agenda for future, autonomy.

1 Introduction

Universities are supposed to be the leading institutions in academics and research. They are expected to transmit knowledge and provide training for various professions and become leaders in creating new knowledge through basic and applied research. The contemporary universities stand at the centre of social expectations. The most important institution in the complex process of knowledge creation and distribution, it not only serves as home to most of the basic sciences but also to the complex system of journals, books, and databases that communicate knowledge worldwide. Universities are key providers of training in an ever-growing number of specializations. Universities have also taken on a political function in society where they often serve as centres of political thought, and sometimes of action, and they train those who become members of the political elite. At the same time, academe is faced with unprecedented challenges, stemming in large part from a decline in resources made available to higher education. After almost a dramatic expansion worldwide, universities in many countries are now forced to cut back on expenditures, and in some cases to downsize. The unwritten pact between society and university education that provided expanding resources in return for greater access for students as well as researchers and service to society has broken down, which at this moment demands some retrospection before setting for future agenda in this regard. This paper discusses some of the innovative challenges of university education in

R.J. Howlett et al. (Eds.): *Innovation through Knowledge Transfer 2012*, SIST 18, pp. 245–253.
DOI: 10.1007/978-3-642-34219-6_27

country like India. Issues such as autonomy and accountability, the role of research and teaching, reform and the curriculum, and the implications of the massive expansion that has characterized universities in most countries are of primary concern here.

2 Assessing the Indian Higher Education Network

The challenges of higher education appear to stem from the failure of the nation's socio-economic structure. Breakdown to dissociate jobs from degrees has put immoral pressure on institutions of higher learning including universities. Given the vertical number of students that most universities in the country have to deal with, it comes as no surprise that these institutions find their resources insufficient to provide quality education. Institutes of higher learning have three major functions: to disseminate knowledge and develop learning; to generate knowledge through original research; and to develop and promote technical knowledge, skills, and products. A large number of universities and institutes are not delivering quality output partly due to overstressed resources and partly due to complete lack of vision. With very thin percentage of GNP devoted to this sector, there is a greater need for higher education to deliver a high quality of output. One leading suggestion, already in the process of implementation, is to make higher education increasingly self-financing. This would not overly burden quality institutions such as the Indian Institutes of Technology (IITs), the Indian Institutes of Management (IIMs), and the Indian Institute of Science (IIS). This would also enable some of the premier colleges in the top central universities to become self-reliant. However, the majority of institutes of higher education that are currently not meeting the aims of higher learning cited above, are likely to face a resource crunch; and more problems might emerge as these poor quality institutions are forced to further cut corners. Unfortunately, since the national psyche is oriented towards the acquisition of degrees, these institutions will probably continue to function.

Perhaps more dire is the fate of areas of learning that do not directly relate to the marketplace. Student loans and collaboration with industry are two important means to generate finances for higher education. Student loans imply repayment and hence the link between an academic degree and employability. Similarly, industry is likely to support institutions and courses that have direct economic benefits. This leads to a possibility that the liberal arts and humanities may be regarded as disadvantaged courses. There are areas where industry-institute collaborations are possible and are valuable. These could include general financial support for the institute, research grants, chairs and endowments, donations of equipment, and participation in curriculum design. In addition, consultancy and the results of research and development are valuable sources of financing. Finally, extension services, student, faculty, and industry exchanges, and joint programmes are possible areas of collaboration. In addition to the issue of financing higher education is the issue of the relevance and role of this education. Curriculum structures, staff recruitment, training, and management of higher education, are areas that have not been sufficiently addressed. Efforts are being made to maintain the status quo and remove academic matters from the debate on higher education. Perhaps this anomaly will have the greatest ramifications in the future

3 Change and Reform in Higher Education

The demands placed on institutions of higher education to accommodate larger numbers of students and to serve expanding functions resulted in reforms in higher education in many countries. Much debate has taken place concerning higher education reform in the last 20-25 years and a significant amount of change did take place. It is possible to identify several important factors that contributed both to the debate and to the changes that took place. Without question, the unprecedented student unrest of the period contributed to a sense of disarray in higher education. Deteriorating academic standards that are the outcome of the rapid expansion in part precipitated the unrest. In a few instances, students demanded far-reaching reforms in higher education, although they did not propose specific changes. Students frequently demanded an end to the rigidly hierarchical organization of the traditional university, and major reforms are being made in this respect. At the same time, the walls of the traditional academic disciplines were broken down by various plans for inter-disciplinary teaching and research. Reform was greatest in several very traditional academic systems. Some universities have completely transformed in the most far-reaching of the reform movements. Among changes are a democratising of decision making, decentralizing the universities, expanding higher education to previously underserved parts of the country, providing for inter-disciplinary teaching and research, and vocationalizing the curriculum.

Vocationalization has been an important trend in higher education change in the past three decades. Throughout the world, there is a conviction that the university curriculum must provide relevant training for a variety of increasingly complex jobs. The traditional notion that higher education should consist of liberal non-vocational studies for elites or provide a broad but unfocused curriculum has been widely criticized for lacking "relevance" to the needs of contemporary students. Students, worried about obtaining remunerative employment, have pressed the universities to be more focused. Employers have also demanded that the curriculum become more directly relevant to their needs. Enrolments in the social sciences and humanities have declined because these fields are not considered vocationally relevant. Curricular vocationalism is linked to another key worldwide trend in higher education: the increasingly close relationship between universities and industry. Industrial firms have sought to ensure that the skills they need are incorporated into the curriculum. This trend also has implications for academic research, since many university-industry relationships are focused largely on research. Industries have established formal linkages and research partnerships with universities in order to obtain help with research in which they are interested.

University-industry relations have become crucial for higher education in many respects. Technical arrangements with regard to patents, confidentiality of research findings, and other fiscal matters have assumed importance. Critics also have pointed out that the nature of research in higher education may be altered by these new relationships, as industrial firms are not generally interested in basic research. University-based research, which has traditionally been oriented toward basic research, may be increasingly skewed to applied and profit-making topics. There has also been some discussion of the orientation of research, where broader public policy matters may conflict with the needs of corporations. Specific funding arrangements have also been questioned. Pressure to serve the immediate needs of society and particularly the training and research requirements of industry is currently a key concern for universities, one that has implications for the organization of the curriculum, the nature and scope of research, and the traditional relationship between the university and society. Debates concerning the appropriate relationship between higher education and industry are likely to continue, as universities come under even stronger pressure to provide direct service to the economy.

Universities have traditionally claimed significant autonomy for themselves. The traditional idea of academic governance stresses autonomy, and universities have tried to insulate themselves from direct control by external agencies. However, as universities expanded and become more expensive, there has been immense pressure by those providing funds for higher education--mainly governments--to expect accountability from universities. The conflict between autonomy and accountability has been one of the flashpoints of controversy in recent years. Without exception autonomy has been limited, and new administrative structures have been put into place especially in new universities that are coming in the recent times. Despite the varied pressures on higher educational institutions for change and the significant reforms that have taken place in the past two-three decades, there have been few structural alterations in universities like curricula have been altered, expansion has taken place, and there have been continuing debates concerning accountability and autonomy, but universities, as institutions have not changed significantly.

This brief background discussion may attempt to provide us the immediate future agenda that universities in India are essentially required for further deliberation and application.

4 The Agenda for Immediate Future

The University in modern society is a durable institution. It has maintained key elements of the historical university model from which it evolved over many centuries. At the same time, it has successfully evolved to serve the needs of societies during a period of tremendous social change. There has been a convergence of both ideas and institutional patterns and practices in Indian higher education in India and it is the matter of fact that universities have been decisive in the development and internationalisation of academics and scholarship. Despite remarkable institutional stability over time, universities have changed and have been subjected to immense pressures. Many of the changes chronicled here are the result of great external pressure and were instituted despite opposition from within the institution. Some have argued that the university has lost its soul. Others have claimed that the university is irresponsible because it uses public funds and does not always conform to the direct needs of industry and government. Pressure from governmental authorities, casual students, or external constituencies has all placed great strains on academic institutions.

In an exceptional good economic growth of the nation, higher education has assumed an increasingly central role in virtually all-modern societies. While growth may continue, the dramatic expansion has likely to come to an end. It is unlikely that the position of the university as the most important institution for training in virtually all of the top-level occupations in modern society will be weakened, although other institutions have become involved in training in some fields. The university's research role is more problematical because of the fiscal pressures of recent years. There is no other institution that can undertake basic research, but at the same time the consensus that has supported university-based basic research has weakened. The challenges facing universities are, nonetheless, significant. The following issues are among those that will be of concern in the coming immediate future and beyond:

a) **Access and Adaptation:** Although in a few states in India, access to post-secondary education has been provided to virtually all segments of the population, in most states a continuing unmet demand exists for higher education. Progress toward broadening the social class base of higher education has slowed and, in many industrialized states, stopped during post-1991. Limited funds and a desire for "efficient" allocation of scarce post-secondary resources will come into direct conflict with demands for access. Demands for access by previously disenfranchised groups will continue to place great pressure on higher education. In many countries, racial, ethnic, or religious minorities play a role in shaping higher education policy. Issues of access will be among the most controversial in debates concerning higher education.

b) **Administration, Accountability, and Governance:** As academic institutions become larger and more complex, there is increasing pressure for a greater degree of professional administration. At the same time, the traditional forms of academic governance are increasingly criticized--not only because they are unwieldy but also because, in large and bureaucratic institutions, they are inefficient. The administration of higher education will increasingly become a profession, much as it is in the developed countries. This means that an "administrative estate" will be established in many universities where it does not now exist. Demands for accountability are growing and will cause academic institutions considerable difficulty. As academic budgets expand, there are inevitable demands to monitor and control expenditures. At present, no general agreement exists concerning the appropriate level of governmental involvement in higher education. The challenge will be to ensure that the traditional and valuable patterns of faculty control over governance and the basic academic decisions in universities are maintained in a complex and bureaucratic environment.

c) **Knowledge Creation and Dissemination:** Research is a central part of the mission of many universities. Key decisions concerning the control and funding of research, the relationship of research to the broader curriculum and teaching, the uses made of university-based research, and related issues will be in contention. Further, the system of knowledge dissemination, including journals and books and the new computer-based data systems, is rapidly changing. How will traditional means of communication, such as journals, survive in this new climate? How will the scientific system avoid being overwhelmed by the proliferation of data? Who will pay for the costs of knowledge dissemination? While the technological means for rapid knowledge dissemination are available, issues of control and ownership, the appropriate use of databases, problems of maintaining quality standards in databases, and related questions are very important. It is possible that the new technologies will lead to increased centralization rather than wider access. It is also possible that libraries and other users of knowledge will be overwhelmed both by the cost of obtaining new material and by the flow of knowledge. The knowledge producers currently constitute a kind of cartel of information, dominating not only the creation of knowledge but also most of the major channels of distribution. Simply increasing the amount of research and creating new databases will not ensure a more equal and accessible knowledge system. Academic institutions are at the centre, but publishers, copyright authorities, financiers of research, and others are also necessarily involved.

d) **The Academic Profession:** The professoriate has found itself under great pressure in recent years. Demands for accountability, increased bureaucratisation of institutions, financial constraints in many countries, and an increasingly diverse student body have all challenged the professoriate. In most universities in India, a combination of fiscal problems and demographic factors led to a stagnating profession. Now, demographic factors and a modest upturn in enrolments are beginning to turn surpluses into shortages. In a newly industrializing country like India, the professoriate has to significantly improve in its status, remuneration, and working conditions in coming times. Overall, the professoriate will face severe problems as academic institutions change in the coming period. Maintaining autonomy, academic freedom, and a commitment to the traditional goals of the university will be difficult. In the university system, it will be hard to lure the "best and brightest" into academe in a period when positions are again relatively plentiful, for in many fields, academic salaries have not kept pace with the private sector and the traditional academic life-style has deteriorated. The pressure on the professoriate not only to teach and do research but also to attract external grants, do consulting, and the like is great. The challenge will be to create a fully autonomous academic profession in a context in which traditions of research and academic freedom are only now developing. The difficulties faced by the universities are perhaps at the peak to maintain a viable academic culture under deteriorating conditions.

e) **Private Resources and Public Responsibility:** There has been a growing emphasis on increasing the role of the private sector in higher education. One of the most direct manifestations of this trend is the role of the private sector in funding and directing university academic and research. In recent times, private academic institutions have expanded, or new ones have been established in India and the students are paying an increasing share of the cost of their education as a result of tuition and fee increases and through loan programs. Governments try to limit their expenditures on post-secondary education while at the same time recognizing that the functions of universities are important. Privatisation has been the means of achieving this broad policy goal. Inevitably, decisions concerning academic developments will move increasingly to the private sector, with the possibility that broader public goals may be ignored. Whether private interests will support the traditional functions of universities, including academic freedom, basic research, and a pattern of governance that leaves the professorate in control, is unclear. Private initiatives in higher education will bring a change in values and orientations. It is not clear that these values will be in the long-term best interests of the university system.

f) **Diversification and Stratification:** While diversification--the establishing of new post-secondary institutions to meet diverse needs is by no means an entirely unprecedented phenomenon, it is a trend that has been of primary importance and will continue to re-shape the academic system. In recent years, the establishment of research institutions, community colleges, polytechnics, and other academic institutions designed to meet specialized needs and serve specific populations has been a primary characteristic of growth. At the same time, the academic system has become more stratified--individuals within one sector of the system, find it difficult to move to a different sector. There is often a high correlation between social class (and other variables) and selection to a particular sector of the system. To some extent, the reluctance of the traditional universities to change is responsible for some of the diversification. Perhaps more important has been the belief that it is efficient and probably less expensive to establish new limited-function institutions. An element of diversification is the inclusion of larger numbers of women and other previously disenfranchised segments of the population. Women now constitute over 50 percent of the student population in many disciplines of higher education in India and more than 50 percent in certain emerging disciplines of academics. In many universities, students from lower socio-economic groups and racial and ethnic minorities are entering post-secondary institutions in significant numbers. This diversification will also be a challenge for the coming times.

g) Economic Disparities: There are substantial inequalities among the students admitted in Indian universities and these inequalities will likely to grow further with the introduction of privatisation in higher education. The top universities generally have the resources to play a leading role in scientific research--in a context in which it is increasingly expensive to keep up with the expansion of knowledge. At the same time, other universities simply cannot cope with the continuing pressure for increased enrolments, combined with budgetary constraints and, in some cases, fiscal disasters. In the middle are academic institutions in India, where significant academic progress has been taking place and it is still not very clear that where these universities are going especially against the underlying spirit of university concept. Thus, the economic prospects for post-secondary education, all-inclusive, are quite mixed and that requires immediate attention to look in to the matter on merit basis.

5 Conclusion

Universities reveal a common culture and reality. In many basic ways, there is a convergence of institutional models and norms. At the same time, there are significant national differences that will continue to affect the development of academic systems and institutions. It is unlikely that the basic structures of academic institutions will change dramatically. The common academic model will survive, although administrative structures grow stronger, and the traditional power of the faculty has diminished. Open universities and other distance education institutions have emerged, and may provide new institutional arrangements. Efforts to save money may yield further organizational changes as well. Unanticipated change is also possible. For example, while conditions for the emergence of significant student movements do not seem likely at the present time, circumstances may change. The circumstances facing universities in modern times are not, in general, favourable. The realities of higher education as a "mature industry" with stable rather than growing resources will affect not only the funds available for post-secondary education but practices within academic institutions. Accountability, the impact of technologies, and the other forces discussed in this paper will all affect colleges and universities. Patterns will, of course, vary from institution to institution and some academic systems, especially those in the newly established, will continue to grow faster than the traditional. In coming time, the universities will be affected by significant political and economic change, the coming decades will be ones of reconstruction in the field of higher education in India. Wide-reaching, the present decade may result major challenges for Indian higher education so as to make it suitable to nation which is an economic-power in making.

References

1. Agarwal, P.: Higher Education – I. From Kothari Commission to Pitroda Commission. Economic and Political Weekly (February 17, 2007)
2. Chadha, G.K., Bhushan, S., Murlidhar, V.: Teachers in universities and colleges – current status regarding availability and service conditions. Higher Education in India: Issues Related to Expansion, Inclusiveness, Quality and Finance. UGC Report, University Grants Commission, New Delhi, pp. 202–213 (2008)
3. Chandrashekar, T.K.: Brain drain and innovation – Convocation Address. University News 46(10), 14–16 (2008)
4. Das, G., Dikshit, H.P., Mehta, G.: A Report on National Assessment and Accreditation Council. Submitted to the Ministry of Human Resource Development, Government of India (2005)
5. Dongaonkar, D.: Missing Links in Education System in India, AIU Occasional paper, p. 13 (2006)
6. Gnanam, A.: National Knowledge Commission: The third empire. University News 45(48), 21–26 (2007)
7. Govt. of India: Eleventh Five-year Plan Document (2007-12). Planning Commission of India, New Delhi (2008)
8. Govt. of India: Selected Educational Statistics (2005-06). Ministry of Human Resource Development, New Delhi (2008)
9. National Policy on Education. Ministry of Human Resource Development, Govt. of India, New Delhi (1986)
10. NCERT: Education and National Development, Report of the Education Commission (1964-1966), Govt. of India, New Delhi (1971)
11. Thorat, S.: Higher Education in India Emerging Issues Related to Access, Inclusiveness and Quality. Nehru Memorial Lecture University of Mumbai, Mumbai (2006)

Author Index